THE EXPLORING WORD

CREATIVE DISCIPLINES IN
THE EDUCATION OF
TEACHERS OF ENGLISH

BY

DAVID HOLBROOK, M.A.

Sometime Fellow of King's College, Cambridge

CAMBRIDGE
AT THE UNIVERSITY PRESS
1967

420.712
H 724

Published by the Syndics of the Cambridge University Press
Bentley House, 200 Euston Road, London, N.W.1
American Branch: 32 East 57th Street, New York, N.Y. 10022

Library of Congress Catalogue Card Number: 66-24114

The completion of this book was made possible by the
grant of a Senior Research Fellowship to the author by
the Leverhulme Trust in 1964–5

Printed in Great Britain
at the University Printing House, Cambridge
(Brooke Crutchley, University Printer)

CONTENTS

For

DENYS THOMPSON

LOVE AND MIND

I

SUBJECTIVE AND OBJECTIVE DISCIPLINES

This book has a simple purpose, which is to set out to define the disciplines of teaching English, in order to suggest ways in which the education of teachers can be improved.

I am particularly concerned with the education of those who are going to teach English in secondary modern and comprehensive schools, to less academic children. But this is not a book which belongs to 'education' as a subject, nor to 'method' as split off from 'English', in the unreal way in which such things tend to become divided in departments and colleges of education. English cannot be divided from the method of teaching it, since even in our exploration of how we use words—we are forced to use words. We need to move continually to and fro, between our own use of language and the child's, and that of the written book: we can never stand outside the word as a medium.

The experience of English in education for teaching (as I shall argue) is itself an experience of method which counts more than any number of lectures on method. If their experience of English at college is bad, then young teachers will arrive in the classroom unready, without the proper 'feel' for the subtle processes of using language well and creatively. They may then fall back on repeating by example what their own teachers did at school a decade before, despite all their instruction in 'method'.

Valuable changes in English teaching in recent years have made it important that this unconscious kind of regression should not happen. Where teachers relapse into out-of-date methods they reveal a failure of training and an ignorance of the essential disciplines of their subject. Through the work of Denys Thompson, Marjorie Hourd, and many other pioneers, English teaching since 1930 has become more meaningful and satisfying. Teaching English in the new ways can be exciting, but there are still many resistances to change—and the shortage of teachers itself contributes. Improvement depends upon training teachers well enough to give them energy and confidence to develop their own

3

independent approach to English. This requires a good enough experience of the proper disciplines in college work for them to be able to recognise dull and bad traditions when they find them, and to know how to replace them.

We begin with the advantage that, in this subject, the business of responding well to literature, of being creative with words, and of becoming literate, is a natural process in adult and child. Literacy is gratefully entered into by most individuals because it brings great satisfactions, enlarges perception, and helps to bring order and beauty to the inner world. There are also other natural processes in adult and child, upon which education depends: so we begin with many benefits and endowments. One is that the child needs to work symbolically with great energy on 'inward' problems, to develop his capacities to deal with the world, to get on with other people, and to establish his identity. The adult, on his part and in adult ways, though he may have completed most of the processes whereby the personality is developed in childhood, needs to go on working at these poetic processes to maintain an adequate sense of identity, a degree of inward order, and outward effectiveness. Language, particularly in its subjective aspects, plays a large part in this; so, an adult should be able to feel an affinity between his own needs to explore experience in words, and those of the child. But the adult is also naturally endowed to help foster these processes in the child, to greater or lesser degree, while the child is naturally inclined to 'use' an adult for his own inward purposes of learning about the world and himself. We take part in this mutual affair because of our paternal (or maternal) and social feelings. The children take part in it because, being animulae, they still need the presence of some 'loved' adult to help them work towards independence—by being there to 'receive' their constructive efforts.

Education, especially the education of literacy, creativity and response to works of the imagination, is a natural subjective process, largely intuitive. It is also a process to do with love, with giving and receiving, and with sympathy and insight. Not everyone can succeed as a teacher. But every young woman who is successful in her first teaching job has taken to this natural fostering process as a duck takes to water *because she is a duck*. She is not a good teacher because she has a file of lecture notes on psychology and 'educational theory', or a B.Ed., or a pass at 'advanced main level'.

I find for my own part that young students have natural gifts, can quickly and naturally establish good relationships with their pupils, and get rewarding creative achievements in English from them, early on in their work as teachers in school. They can do this because they are not so very far from their own childhood, and so are aware (inexplicitly) of what children need to do, by way of exploration, in phantasy, of inward experience. They are also very flexible and want to use their paternal or maternal powers, out of sheer excitement at being newly adult. Sometimes students are unsure whether or not the results they have obtained have validity. They are often perplexed because, though they can teach well, they cannot yet make the process explicit: they cannot easily, for instance, justify their approach, if challenged, to a hostile authority. Yet in themselves they know what they are doing to be right: what they sometimes need is backing, because of insecurity of conviction, and lack of experience. But this problem would seem to be worse sometimes *because* they have been trained in abstract 'theories of child development': struggling to fit their work to a conscious theoretical scheme they become confused and cannot 'allow' how marvellous the intuitive process has been.

The starting-point of this book is this: if we examine the experience of an English teacher in the classroom and ask what it is he is really doing, we may go on to ask, does his training provide the best preparation for this? We are now being urged to plan for some student teachers to be given a B.Ed. and to take a particularly exacting course. The universities may become increasingly responsible for teacher training. What will the effect of these innovations be? Will they help to make teacher training more realistic and efficient?

I began to interest myself in teacher training having come from the secondary modern classroom, from adult education, and from teacher 'refresher' work. I have held seminars with graduate students from a department of education, college of education students, and interested undergraduates, for five years, and have been external examiner for four colleges of education. I have visited many colleges and have done periods of intensive teaching at some. I found, to my dismay, a huge gap between what a young teacher needs in the classroom, and what we provide in training him. I believe that this gap could widen, even because of the new developments towards B.Ed., through the consequences of 'reorganization', under the pressure of false disciplines, over-intellectual substitutes for the real work, and misguided pretensions. This could happen unless the universities and other governing authorities become able and willing to accept, to study, to value and respect the essential creative disciplines of whole being on which teaching depends. On our side, I am sure, as teachers, we must uphold the significance of intuition and 'touch', and resist ignorance, misunderstanding and such impulses of intellectual hostility as threaten the great creative movement in English teaching, despite its new 'official' endorsement by such bodies as the Schools Council. If there are to be more thorough and exacting disciplines, these must be the real disciplines of the exploring word.

2

TEACHING COMES BY NATURE

Sometimes, when things go wrong, in gropings after formulations for the B.Ed., no one individual seems to be to blame for what happens: false disciplines seem to breed themselves out of the air. Perhaps it is at times because of ignorance—at times because of lack of work done.* As external examiner I have noticed on at least two occasions that suddenly the year's syllabuses and question papers at colleges of education have become extraordinarily academic—full of topics of minor scholarly interest—high-sounding but vacant generalisations, references to little-known works, requirements for baffling 'links' to be made with philosophy and aesthetics (Burke on *The Sublime*, the English Moralists, Satire and Epic in English poetry and so forth). It was not so much that the syllabuses had gone beyond the capacities of the kind of student who goes to a college of education. The students *could* have tackled them. But the syllabuses had become irrelevant to the central concern in English teaching, and became appropriate rather to the training of a textual scholar or university lecturer, both of which are very specialised professions, and as remote from the life of a school as could be. To tackle such abstruse syllabuses could only have been a waste of time for student teachers who have only three years to learn a complex discipline. They needed something much more whole, direct and immediate—a month on poems of Edward Thomas, say, or on books like *The Shadow Line*, rather than on 'The Gothick Novel', or the 'Romantic Revival' (leaving barely time to read a single poem), or half-digested 'information about' philosophical movements (with no time for one philosophical thought).

When I asked why the syllabuses had become so abstract, dry, and academic, the answer came that 'the university expects us to raise standards because of the new B.Ed.'. The college lecturers perhaps saw in this new development a chance to feel 'like university staff', to enhance their status by feeling they were stretching towards a university

---

* It is interesting to note that the report of a working party of teachers on *English in the C.S.E.* has been taken over lock, stock and barrel by some regional examining boards. Naturally, we are pleased: but I have no doubt this happened simply because so few other people had done any work on the subject.

'level'. It may be good for what we now call 'schools of education' to feel part of the university atmosphere: but why do the new
developments have to take this particularly dry, niggling direction in
the dusty byways of Eng. Lit.? Must the approach to literature for
'higher standards' be necessarily so useless? Must linguistics become the
menace grammar once was? Are the true disciplines of response to
literature so much 'idleness and sloth' that we must find a substitute?
This is not the experience of the literary critic!

In one centre I found a way out when I brought the professor and the
lecturer in the college face to face. The professor confessed at once that,
far from any 'requirements' having been issued, the Board of Studies
had few ideas at all, about B.Ed. or, indeed, about disciplines in general
for English in education. What did I think? Had I any suggestions?
But in another place I found the college English lecturer who had
questioned the standard had been asked if he 'dared to criticise the
university'. Elsewhere again, the college lecturers, primed with what
they felt to be appropriately academic syllabuses, were confronted by
a university committee who *were* aware of the true disciplines of English teaching: 'Where is your creative writing?'—'Where's Mark
Twain?'—'Where's Dickens?' Here, the boot was on the proper foot.
But in both places, the spectres bred out of the air—out of the habit of
feeling that 'they', 'the university' (synonymous with the 'university
examining board', of course, in many people's minds) expect '*higher*'
standards—which means something more baffling, more 'scholarly',
evidently '*harder*' than before (often trying to be more 'objective').
Yet it may be further still from the relevant disciplines of whole
response to words.

No doubt sometimes a Board is persuaded that this *was* what it
wanted: but the menace, I am sure, begins to form in the first place, in
the air between the two parties, sometimes out of a fear of having
ignorance exposed, or some other such panic. The imposition of such
falsely pretentious syllabuses masks some absence of a real understanding
of the essential creative disciplines of the word, and of adequate concepts of what 'learning' may be. Sometimes mere Sunday newspaper
values are imposed instead, for the same reasons.

So a false 'modernity', or an insistence on irrelevant academic effort
under the guise of 'rigorous disciplines', can both form out of the unconscious processes—pretentious because self-deceiving—by which we
'educated' people tend to prop up our identities and disguise our

ignorance and weakness by intellectual defences. All syllabuses tend to be window-dressing in this way, even from the best intentions. No student can really absorb, thoroughly, more than ten major creative works in the year—fifteen at most: but the syllabus will suggest thirty. We may ourselves be deeply embarrassed to be asked to give an adequate account that is our own of Wordsworth's *Michael* or *Wuthering Heights* or *The Rainbow*. Even if we could do it, do we want to talk about such uncertainties as love and hate, and vindicate our position by close textual analysis? So we take refuge in remarks about 'Satire' or 'the Romantic Poets' or 'the Brontës', about which almost any vague generalisation is safe enough. Thus we avoid the pain of being obliged to adjust ourselves and to have it out with our students in that exhausting way all over again. As one discovers in one's own work, the syllabus is a piece of show which, when done, will satisfy all kinds of people—directors of education, committees, even students. A syllabus will even secure grants. Once it is printed or duplicated it looks as if the work has already been done. No one need fear any longer the unhappy, insecure and intangible human process that sharing the reading of a novel or a poem is, in the classroom, with all that discomfort of feeling exhilarated or depressed, and with the risk of feeling, sometimes, 'they haven't got it'. The duplicated or printed syllabus makes it seem that the students are already in possession of all these works—although they may never read any of them, and have really only 'had lectures on' them.

In a sense, any syllabus is a sheer lie—an orderly seeming abstraction, like the diagram of a battle, while the human experience itself is always pretty chaotic. Sometimes the most able student's possession of works in his syllabus—like some of yours or mine—is partial, haphazard, and distorted, no more than a few scraps of impressions, odd fading memories circulating in the dynamic processes of his inward personality. Our disciplines have to do with improving capacities to take in, and to articulate impressions from reading: but in a recognition that there is no question of tidying the process up. There is no place where perfect readings and judgements exist: we only deal with individual partial possessions—though, of course, these can be improved.

The way out is to look closely and honestly at the incomplete and personal thing our work essentially must be, and accept what it has to do with: largely 'precipitates from the memory'. Then we shall discover, in the living experience, elements not taken into account by the syllabus.

Indeed, we may find elements which force us to see that the syllabus as we have mapped it must be scrapped—because as an abstract it has lost all meaningful touch with the processes it was originally supposed to foster.

Such a drastic revision of content and method needs to accompany present-day expansion in the training of English teachers. The schools have been moving ahead at a great pace towards a deeper understanding of what children need. Meanwhile most colleges of education are far behind, and at a time of expansion need to become more efficient in the right kind of way. There is, I think, a danger of superficial 'reforms' and a false inclination to substitute external changes in organisation for much-needed changes in content (again, to pretend that 'inward' problems can be solved by 'outward' management, as with the hasty change to comprehensive education).

We can, certainly, ask some fundamental questions about the content of education courses. How much, for instance, do lectures on psychology, the history, aims and methods of education, and the rest, contribute to the student teacher's capacities? As an educational experience themselves, cannot such lectures sometimes be irrelevant and even a bad influence? They imply by example that the teaching process is more rational, more under control, more 'abstract', more 'objective' than it ever can really be—and that education is taking in, taking notes. Such questions have as yet hardly begun to be asked. My scepticism here should not be taken to mean that intuitive powers cannot be improved by conscious effort and awareness, such as seminars in child development or in theories of education can promote.

What I shall suggest here is a syllabus based on *experience*—based essentially on work with children and the observation of children, followed by seminars with those who have shared these experiences, who have a little more experience of life, and who know a little more than the student. The controlling focus will be the meaning of children's words and the light this casts on their need for literature.

When I speak of those who 'know a little more' I do not necessarily mean those having abstract knowledge such as is gained from experimental psychology, but those who possess and can articulate the 'whole' knowledge gained by experience of children, and from sources which offer insight (such as painting or writing poetry) into the child's inward processes. Such sources are not confined to empirical 'outward' study, such as educational research mostly fosters, but include the insights gained by projective identifying, from observation, from the study of living children and from analysis of the content and poetic meaning of children's work. Much revolutionary work in English in recent decades has come about from such approaches—and from insights gained by child therapy and psychoanalysis, which deal with those areas from which poetry comes too. Some of the best developments have been made in the work of people who are 'untrained' in the traditional sense, and who had no degrees—but did have the intuitive gift of being able to make touch with children and their needs. In so far as I comment on organisation, I shall urge attention to the students' experience of education, because this is the way he learns the art of teaching—not only by 'knowing about'. The student should be *doing*, and *doing with children*: and then making his experience and observations articulate, and discussing them. Finally, I hope to suggest how in English teaching collaborative exploration of the creative process between child and teacher can be fostered by a relevant use of the resources of English literature, by drawing on relevant artefacts of our civilisation—the record of explorations of experience made and left by the dead, from which we can benefit. To take possession of this legacy it is futile to seek to 'cover the ground' as by 'survey courses'. What we must try to do is to give an appetite for the poetic and a capacity to read well, in such a way as to serve a teacher throughout life—enabling him to continue his own exploration from the indications and bearings given him by his English course at college. That is, we should try to foster a dynamic rather than the possession of a rag-bag of scraps of information.

At the same time, it needs to be emphasised that the student teacher is also a young growing person who needs to make her own possession of her language and her civilisation: and that in this her relationship with her tutors will be an experience which will be significant in itself, in terms of its degrees of civilised relationship. Obviously, unless the whole life of a college matches the content of the culture of the lecture-room, some kind of dissociation will be felt—and repeated in the student's work in schools in the outside world. I wish I had the experience to discuss what kind of civilised community a college should be from the 'inside': but the connection between the subject of English and civilised living in the institution will be implied in much of what I say.*

In such a view of teacher training, of course, pretensions can have no

---

* At one college the men students sleep in dormitories with partitions only between 'rooms'. Such inadequacy of provision for privacy for young adults is surely not civilised enough?

place: so, the ground is difficult and disturbing, not least because we need to arrange things in such a way as to prevent us from deceiving ourselves. We may never do as much for a student (or a pupil in school) as we may have assumed. At the same time everything depends upon them doing all for themselves—and we meanwhile doing everything for ourselves, to try to keep our own work creative and realistic. This means that college teaching staff must teach and continue 'doing', themselves. They should be given opportunities to continue to teach in school and remain in touch with teachers and children. Teaching must be defended for what it is—essentially an intuitive art—disturbing as this truth may be.

Since education deals with inward processes, and our experience of it is inevitably the experience of persons, it can never be other than subjective. No study of the partial functions of individuals is ever a substitute for the whole picture of this whole experience: and the use made of much 'objective' data is itself often so subjective that the supposedly careful 'objectivity' of the data is rendered invalid anyway. How is one to convey to the 'objectively minded'—scientists on college committees, say—the nature of the real disciplines of literary criticism and their application to the creative writing of children? Or how poetry helps one to understand children's personality problems? Unless one is met half-way by intelligent understanding of human nature, these complex vindications of subjective disciplines will sound lame beside talk of 'rigorous objective research'.

Yet if one examines many pieces of educational research and ignores the way they are presented to resemble a science, they may often be found in the end to be dealing with equally subjective factors, since they are dealing with human beings and their inward life, even if under a heavy (and sometimes disqualifying) disguise of graphs and tables. Even justified measurement of, say, the time taken over simple learning processes, can never demonstrate any conclusions as to *why* essentially one learning process is more effective than another: this can never be anything other than subjective. The pseudo-scientific jargon and method, and the new 'technology' of research, tend to imply a denial that this is so—possibly because of unconscious fears of the subjective, of that area of darkness which speaks of our unknown selves, and of those aspects of life which can never be subject to logic, reason and 'control'—and yet can threaten our whole stability and reality sense at times. Often 'researchers' seem to be doing so more than trying to erect a model of grossly oversimplified abstractions, controllable to the degree of being dead, against the teeming creative complexities of whole living creatures. Intellectually, they can be exposed by such a doggedly sceptical intelligence as Professor Bantock's (see *Education and Values*, pp. 153–74, on *Educational Research: A Criticism*).* Their enemy is the mind which thrives on the subjective and embraces intuition—the mind that creates and thinks in whole terms.†

There are other important ways of exploring the educational experience, by projective identification, and careful observation. A man sitting gazing at a playground, or watching a baby with a spatula (there is a marvellous account of this by D. W. Winnicott, *Collected Papers*, p. 46), reading children's poetry, or simply sitting and thinking over his teaching, may not be doing 'research': but he can be discovering a great deal, subjectively, about the nature of children. It is too readily assumed nowadays that 'research' must be dealing with externals: it must have questionnaires, data, pigeonholes and staff. Yet while much money is expended on this card-work there are hundreds of teachers who need time off to reflect quietly on their work—who would perhaps, if they had time, do much to illuminate our experience, by learning to make theirs articulate.

Yet in many places it is psychometric 'research' which exerts a major influence on the training of students for what is primarily an intuitive art. This struggle between 'objective' disciplines and the essential subjective ones is likely to become exacerbated by expansion, towards 'university status' and B.Ed.—if only because universities understand objective research, while they have much less time and respect for subjective and creative disciplines and for introspection.

The problem needs to be seen, of course, in terms of what all the effort in research, teacher training, and teacher refreshment is directed towards. We do, of course, seek to make education more 'efficient': but nowhere does the impulse towards efficiency more require to be governed by a deep sense of adequate aims, of relevance to human needs, and the needs of society. In so far as there are greater needs than that implied by saying 'the community requires 1,000 dentists' or '10,000 lathe operatives', or 'this university will aim to produce the kind of general purpose student industry requires' (as the vice-chancellor at

* Examining some questions put to subjects in research, Professor Bantock comments, 'One wonders in how many pieces of research the neglect of similar subjective interpretation of the situation is not commented on or even remarked' (p. 174).

† See Appendix B, a note on 'Wholeness'.

13

Norwich put it), even these need to be conceived in whole human terms. Such aims, to be realistic, must recognise that our capacities to deal effectively with the outer world depend upon the degree to which we are able to deal effectively with our subjective world. So, they are the concern of imaginative and sympathetic human insight—the field of poetry, the arts, of care and nurture, and of psychotherapy as a practice. The 'practical' man who supposes aims in education can be dealt with in objective and outward terms is being unrealistic.

Of course rational considerations and trained intelligence are not to be despised, and I am not urging these should always be subordinate to intuition. The point is rather that the intelligence can be too easily undermined or thwarted by the 'unknown self', while the intellect, on the other hand, can be used as a weapon to deny intuitive needs. In a democratic society what we seek to foster is a maturity by which individuals are capable of independent judgement—of making their own choices. This involves inward creativity—and often a conflict between love and hate within—as over a decision whether or not to behave in such a way as to live at someone else's expense (the kind of subjective problem underlying racial issues).

Judgements of this kind involve both conscious intelligence and the whole personality: both can be trained to be sympathetic, guided by insight, and discriminating. A 'whole' training will help prepare a student in these capacities. A training which is dominated by a supposedly 'objective' discipline may not. To ask someone to submit judgement and discrimination in living to the interests of an unobtainable 'objectivity' is thus something of an insult to the fullness of human 'being': it is also often an attempt to deny the inevitable truth that 'we all share the same darkness'—as with the logician who has a complete and orderly scheme, achieved only at the expense of leaving out major elements which cannot be included, and so are denied altogether.

The best kind of educational psychology is that of those such as Professor Cyril Burt whose human judgement leads them to a particular problem. Such psychologists experiment by the rules of their discipline as objectively as they may: but their judgement (essentially based on subjective factors) comes into play again in evaluating the relevance of the empirical study and leads them to their conclusions. As we shall see, the trouble is that too much educational research and training tends to be dominated by those who are obsessed by 'empirical research and controlled observation'. They elevate the means, the instrument, into an end, in their concern with an ideal 'method'. As things are, these impulses tend to be preferred by the universities, because they understand them better than the creative processes of 'whole' learning.

The related problem of teacher training, as it is raised by the proposals for a new B.Ed., was put succinctly by an article in *The Times Educational Supplement* in the issue of 13 November 1964, by Professor W. R. Niblett, Dean of the University of London Institute of Education. Professor Niblett suggests that the candidates for B.Ed. will be distinguished from other undergraduates by the fact that they will already have chosen their vocation, and are likely to be people less interested in abstract theories than in 'actual situations...and questions ...related to action'. He suggests therefore a course 'based on and linked with a growing knowledge of children and schools'. This knowledge, he suggests, is gained by young people from memories of their own schooldays, the reading of books (not least those dealing with schools and young people), actual experience of getting to know children by working with small groups (e.g. in their homes, in children's clubs or play-centres), from visits to schools, and, finally, from teaching practice.

Without an 'exercised capacity for introspection' and this 'widening experience of schools', he says, most B.Ed. students will be unable to 'bring to their study of education—whether of the philosophy, psychology, history or sociology of education—that degree of engagement with life without which these subjects (even when knowledgeably taught) can easily grow bookish and barren...'. This is an important observation. Professor Niblett then discusses what is meant by 'rigorous intellectual disciplines' and concludes that in teaching students another kind of mental activity is relevant: 'there are some kinds of knowledge which can only be received by a mind that *is* relaxed (most of the arts for example), or a mind that is allowed to roam freely...'.

This links up well with Marion Milner's insistence on the need for opportunities for 'reverie'* in education. But despite this important paragraph, I feel that Professor Niblett's approach is in the end still too academic. He ends by discussing 'the interdependence' of aspects of the teaching of education: he sees collaboration in which he foresees '...methodology, for example, being conscious of its dependence on

* Also for 'surrender of the planning conscious intention' and for the state of 'emptiness as a beneficient state before creation' see *On Not Being Able to Paint*, by Marion Milner, pp. xiv and 163 ff., and Appendix B on 'Wholeness', below.

child development, sociology aware of the value-judgements lurking within the pattern of the curriculum and the philosophers able to point out that the priorities decided upon by educational administrators incorporate value-judgements, too'. This sounds like a useful indication of co-operation between subjects. But surely such illuminations are more likely to come—and come with greater depth—in students who have responded to works of literature and imagination, than from explicit intellectual exchange at a superficial level, which is all there is often time for in 'mixed discipline' courses? A student who has possessed the challenge of the question 'What for?' in a whole human experience as explored in a work by E. M. Forster or D. H. Lawrence, is far more alert to 'value-judgements lurking' in the curriculum than most sociologists show themselves to be. Much the same is true of the methodologists, whose basis for learning in experimental psychology so often fails to take into account the whole being and the intrapsychic life, the inward processes of any actual human creature, their inacessibility, the illogical logic, and their creative and dynamic aspects. The complexities of the unconscious roots of development of the personality, in terms of intrapsychic dynamics, from the first phantasy-dominated stages of discovery of the difference between 'me' and 'not-me', are also overlooked in much sociological debate about the 'individual' and the effect upon him of 'society'.

Education cannot be studied without attention to the development of the living personality. We deal with whole persons, not the functions of those fragmentary aspects of persons with which the empirical psychologist deals. Whole experience can only be apprehended by those forms of identification, introspection, insight and judgement that belong to poetry and the creative arts (or psychoanalysis in practice). The mechanics of experimental psychology cannot touch them; and often the impulses of this 'science' are to deny the complex inward processes, out of fear, because they are inaccessible to mensuration, and so belong to a dimension that sometimes makes science uneasy. Besides, in seeking 'objectivity' experimental psychologists sometimes demand of the researcher that he should suspend his own judgement, which would often tell him that what he is dealing with is trivial or partial.

Professor Niblett, however, fails to mention two significant elements of 'education', that must be central to our concern with new standards and disciplines. For one thing, every young teacher has an intuitive capacity (which we take for granted) to foster children's growth and capacities. Abstract information and 'rigour' of intellect is not sufficient to foster the flourishing of this intuitive capacity. 'Intellectual rigour' of the wrong kind could even undermine a student teacher's confidence in his natural creativity and so inhibit it. Secondly, what above all does develop whole powers of intuition, such as we require for teaching, is creative work, and response to creative work. Young people working in clay, or music, and watching the creative work of others (as in mime or drama), or listening to others making music, are intuitively, with that 'relaxed' and 'open' mind, learning about the inward life. To what they gain in these pursuits, they can then more effectively add what they learn from theory: and from this creativity at the centre—as well as their experience of children—they will be able to use and judge Rousseau, Plato or Froebel. Their intellectual preoccupations will be given the touch of that 'life' Professor Niblett desires to keep green, by their exploration of experience in creative terms. This is the best discipline.

But even Professor Niblett's tentative liberal suggestions were too much for some. They were seen as a threat to the claims of experimental psychology by Mr L. B. Birch, editor of the *British Journal of Educational Psychology*, who wrote to protest. Mr Birch represents the prevalent distrust of 'subjectivity' and 'intuition' (which are 'scientific' terms used for the disparagement of 'judgement' and 'insight'):

I wish to protest about the undue emphasis and approval which Professor Niblett places upon the reading of fiction and autobiography (which is often not readily distinguishable from fiction) as an important means of introducing teachers-in-training to the study of children. It is true that this is a very prevalent practice. . . most frequent in those colleges which had no qualified lecturer in educational psychology on their staffs. Where there were lecturers in education with psychological qualifications the study of child development tended to be based upon the professional literature of which there is no real shortage.

One cannot help but conjecture that where the teaching of the psychology of childhood is in the hands of 'amateur' psychologists, the intuitive insights of authors not trained in psychology tends to be preferred to the writings of professional psychologists whose views, based upon empirical research and controlled observation, may be at variance with those found in the writings of fiction. The intuitive genius of Shakespeare portraying the madness of Ophelia may be greatly admired, but one cannot imagine *Hamlet* as a first set book for intending psychiatrists, any more than one can imagine science fiction would be recommended as an introduction to physics.

Note in this letter two aspects: first, the hostility to 'fiction'. As Mr Gradgrind insisted, we are not to fancy. Autobiography is often 'not readily distinguishable from fiction'—and so, suspect of 'subjectivity'. Fiction is not 'true': it is a lower form of life altogether than that kind of knowledge which is 'based upon empirical research and controlled observation'. (In what sense can any deep study of human personality and development ever really be empirical and 'controlled'? Are any of the really important aspects of the personality measurable?) Secondly, reading fiction is referred to as a 'practice', meaning a bad practice—almost a vice. It is almost suggested by Mr Birch's letter that experience of children (or of other human beings) by the normal processes of identification and social contact is itself useless, because it is 'at variance' with the findings of empirical psychologists. Tell that to nursing mothers and ten thousand creative young women in infant schools, who have no explicit idea why their work is so marvellously successful—which is because they are young women with natural gifts for dealing with infants! If mothers were not intuitively able to mother, and able by the exercise of judgement and introspection to improve their performance, we should all be insane: something of the same applies to teachers and the extraordinary way so much education 'works' and works well. There may be 'no shortage' of non-fictional literature about child development, but how much of it is insightful? How much does it really foster our intuitive capacities, or any capacities? What Mr Birch is implicitly denying is the intuition without which nothing in parenthood, upbringing, or education would work at all. Yet it is evidently the basis of all such creative accomplishment.

There are areas of knowledge where 'controlled observation' and 'empirical research' are valuable. But there are many areas which these disciplines cannot touch, not least our dealings with whole persons in relationship—as in the teaching situation. Marjorie Hourd's analysis of children's writing and its creative content, or Dr D. W. Winnicott's analysis of what a baby is doing when he is playing with a spatula, are based on the interpretation of metaphorical and symbolic forms of expression. They deal with not easily accessible areas of inward life: but by the disciplines of interpretation and identification we may come at these. They may be subjective, only in a limited sense empirical, and not 'controlled': but they are more useful than the measurement of partial functions, to our kind of understanding and creative work with children. To understand the processes of the inward life of a child we need not only to watch what he does, but to consider the symbolic content of it—what he means by it. In this there is bound to be much 'subjectivity', whatever control or empiricism we exert: thus, the things of which Mr Birch speaks, and of which professional psychologists write, belong to our disciplines as much as to theirs. In a sense they are using a limiting methodology to invade *our* territory—it is not that we are invading theirs! Good teachers were successful before the psychologists came along: how much has psychometric psychology really done for them?

As for Mr Birch's scathing reference to Shakespeare, one can only reflect gloomily on the attitude it reveals. In seeking to understand the nature of a withdrawal state in a child, or a schizoid state of inability to choose and act, or to understand the part the Oedipus complex plays in our inward life, I can think of no better source of insight than Shakespeare's *Hamlet*. This has little to do with any supposed 'realism' of the madness of Ophelia: Mr Birch shows, in the way he discusses *Hamlet*, his ignorance of how imaginative works enact their meaning. The insights they yield are deeper than those suggested by supposing that Ophelia is a single case-study of a certain kind of madness (or, say, Leontes in *The Winter's Tale* is a study of paranoia). The whole play, as an expanded metaphor, expresses a state of dissociation in the human soul, which was Shakespeare's. It is universal, because it is our continual perplexity, such as many children are faced with in their inner life and which we continue to be faced with as adults—for instance, as when the identity could 'thaw and resolve itself'. It would be possible to measure and 'controlledly' observe a hundred children, while never coming near perceiving or registering such an inward perplexity: yet a teacher who knew *Hamlet* might spot it by sympathy.

Such living problems defy empirical investigation; yet may be explored by introspection and identification—by imagination. Many 'professional' psychologists have tried to touch on such matters—and failed: but then along comes 'an amateur', such as Mark Twain, or Lawrence, or de la Mare—and offers more illumination than a ton of copies of the *British Journal of Educational Psychology*. Or along comes one person with more insight than empirical impulse such as Susan Isaacs, Margaret Ribble, John Bowlby or Winnicott and the whole view of childhood is changed—not merely by 'controlled' 'empirical' observation, but by open-minded sensitive perception and the pondering of experience in an insightful way. This insight is not merely subjective, for it is confirmed by the experience of others, as the whole of psychoanalytical knowledge is embodied in thousands of papers and case-histories, all based on subjective insights. They are based on individual subjective identification and analysis of symbols: but are confirmed by finding universal elements in other human beings, and by the consensus of many observers.

For such imaginative insight there is no substitute in mere empiricism, while literature is often full of such substance.*

Mr Birch's disparagement of the disciplines of 'reading of fiction' as opposed to the superior validity of 'psychometrics' (which of course will seem like a 'real' discipline for the new B.Ed.) sounds convincing, because it seems to offer a security of knowing what one is doing by having an 'external' validation for it. Fortunately for life's potentialities, we can never fully know what we are doing. Mr Birch's parts are not wholes, and his partial view has no claim to dominate the whole of education: the humane subjects, on the other hand, have.

So, when we come to the question of syllabuses and organisation of arts subjects, we not only find the disciplines difficult and difficult to define—we find we are exerting them and defining them against opposition. I have heard recently of a committee suggesting that at one college where the students have been doing sculpture that it would be more valuable for them to study the history of sculpture. As I suggested above, this opposition has unconscious origins. It may be a rationalisation of unconscious defences against the pain of apprehension of the inner world, requiring growth and change. Opposition comes both from those who would prefer the superficiality of the journalistic mind to true creative engagement—often a manifestation of an envy of creativity—and from the scientist who seeks unconsciously to deny inner reality because he wants to subdue the subjective to the objective. F. R. Leavis indicates the nature of this perplexity in the humanities:

...History and English go together in this respect: pondering the nature of the thought proper to either study, and the criteria of quality of thought,

* As for science fiction—who would find that comparison relevant to Shakespeare? Science fiction is itself a schizoid activity—phantasy about an unreal world, in which the emotional truths and realities of this one may be conveniently ignored or denied for the purposes of the writer. Science fiction is, like all other phantasy expression, a projection of inward dynamics in metaphorical form and must be judged as such: the trouble with most science fiction is that it is not very good. Mr Birch's analogy is unfair, since no responsible person concerned with the insight that imaginative work can bring would ever suppose that what he was doing was comparable with using science fiction as an introduction to physics. To use Donne's poetry as a way into problems of personal relationships would be a valid example—and it would be a better (and more 'whole') way into the problem than by Mr Birch's kind of psychology.

one might say of either that it is at the other end of the scale from Mathematics.

Thought in the literary field involves sensibility and value-judgment. Any real literary education entails the development of sensibility. One can't read with someone else's sensibility, and a judgment is one's own, or it is nothing. That is, there can be nothing schematic about processes of thought belonging to literary study, and nothing demonstrable about their conclusions. Yet literary study, in so far as it deserves a place at the university, is essentially concerned with a discipline of thought.

To bring this truth home to a governing body composed predominantly of scientists is virtually impossible. They can't see why the class-list showing shouldn't, on the whole, be taken as an authoritative index of quality. If one at all often (and there is often reason) points out that *A*, who has a II, is in fact alpha-plus—capable, that is, of distinguished original work—while *B*, who has a I, has nothing but the kind of ability that succeeds in journalism or at the B.B.C., one's judgment soon ceases to carry much weight. One must expect to hear as a kind of refrain to one's life that Tutorial Authority has again thrown out the muttered menace, 'The English men don't work hard enough' and one knows for certain that, in the official view, a good supervisor (or Director of Studies) is one who 'gets Firsts'.*

If the colleges of education come closer to the university, they will be surprised to find some energetic opposition from scientists and others, who sooner or later find themselves on planning boards, to the disciplines of English and other imaginative subjects—and to the creative element in teaching itself. These people cling to certain exclusive realities even to the length of denying other obvious realities (such as love or conscience) which the scheme of empirical science cannot include by its very nature and dimension, and which in consequence are hated for their threatening intractability; or ignored.

There is perhaps some justification for the desire of scientists to resist expenditure on chapels (remembering Galileo). But it would seem unbelieveable that there should be scientists who will go so far as to deny the value of English altogether, often showing remarkable undercurrents of hostility as revealed in this news item:

Dr W. Belton, Chairman of the Board of Studies in Science in the Leeds University Institute of Education, explained today what he thought the form of the full year B.Ed. degree recommended in the Robbins Report should take.

Speaking to the conference of the Association of Teachers in Colleges and

* Letter to *The Sunday Times*, 9 August 1964.

Departments of Education, he said that the task ahead was to show that education was a proper university subject and not just a hotch potch. The status of the degree must be right in the eyes of university authorities and subsequent employers. The object was to turn out people who could do a job. People should be able to look back and say they are educationists in the same way that physicists say they are physicists.

The new degree, to give it parity with other degrees, should be general, consisting of education and two sciences, yet retain a broad basis since it is not designed to produce specialists.

Dr Belton suggested that the first year should provide a foundation with courses in education, physical and biological sciences, and the relevant mathematics; and second and third years could combine education with two subjects chosen from physics, chemistry and biology; the final year could combine education with just one of these subjects, leaving room for individual studies where possible.

Dr Belton, touching on a controversial topic, was doubtful of the value of English by itself. This, he felt, should be related to scientific studies instead of being put into a separate compartment. Nor did he see how the new course could be combined with an arts subject. 'People who want to do this sort of thing,' he said, 'ought to go to a university.'

Dr Belton ended with a plea for reductions in the staff–student ratio and improvement in equipment and facilities. Colleges of Education had a strong claim to these improvements, he said, if their courses were to have parity with degrees in universities.*

Here 'parity with degrees in universities' would seem to imply the exclusion of the central discipline. For a scientist to take up such an extreme position reveals a lack of realism—a fundamental inability to see the nature of the teaching process and the basis of science itself in metaphor, and the need for literacy even (or *not least*) for scientists. If a scientist is interested in evidence, he has only to observe a classroom situation at any level to see that education is conducted in the English language, and that learning is a process in complex with personality development in which language has a central function, and whose area is that of inner reality. It ought also to be possible to convey to him in what sense subjective exploration is the gateway to objective awareness. Since these are functions and processes whose creative medium is English, to exclude English from an educational course would be as ridiculous as excluding all training in empirical measurement—or mathematics—from a science course. English is the fundamental sub-

* *The Guardian.*

ject, since language is the medium of all education, and of thought and feeling, and in all its modes is a means to extend and deepen the individual's perception of his inner reality, and to enable him to probe outward reality. Dr Belton is attacking the essential medium of all education: the language of creative exploration and growth.

How can an intelligent scientist come to find himself in such an irrational position? Only surely by some such unconscious denial of the subjective world as I have suggested? No doubt the disciplines of science and creativity can meet; but it must not be in an atmosphere of supposing that the intellectual-ratiocinative disciplines are more 'rigorous' or 'higher', or that objectivity must be predominant, even at the exclusion of subjective disciplines. Science must simply learn to accept the subjectivity it now sometimes tends, out of fear, to deny.

The problem of the relationship between the arts and science becomes acute in the training of teachers, for the essential disciplines of education are to be found in our whole experience of creative processes in the child and in the adult who teaches the child. These disciplines are with the growth and structure of individual personality towards maturity. One means towards the growth of the structure is the word, which moves backwards and forwards as symbol between the self (the ego) and the inward dynamics of the unconscious.

How words relate to phantasy and symbol, and what part they play in the developing stages of human consciousness and personality, is a complex subject which we need to continue to explore. In a valuable paper on symbolism the Kleinian psychoanalyst Dr Hannah Segal suggests that to be able to 'use' symbols in exploring our inward and outward life itself marks an achievement—the completion of a stage. A schizoid patient, she points out, may not be able to distinguish between a word and the thing it represents: only later than the 'schizoid position'* do we become able to see there is a difference between a word used symbolically and what it represents. She also emphasises that the basis of thought and communication is communication with oneself:

The capacity to communicate with oneself by using symbols is, I think, the basis of verbal thinking—which is the capacity to communicate with oneself

---

* Roughly speaking the 'schizoid position' is a stage of growth from which schizoid illnesses arise: it marks the first discovery of the identity, and thus of the 'not-me', too. The next stage is the 'stage of concern' at which the infant becomes perplexed at the possible consequences of hostile phantasies at a stage when he is neither fully sure of the difference between 'me' and 'not-me', nor between phantasy and reality: difficulties here lie at the root of depressive illnesses.

by means of words. Not all internal communication is verbal thinking, but all verbal thinking is an internal communication by means of symbols—words.*

The symbolic communication with oneself is a means to seek order in, and come to terms with, dynamics in the personality which cannot be dealt with except in metaphor, except by symbolic phantasy. Art is thus an activity of resolution of identity, by phases of activity which reinforce the structure of the personality, and organise the content of the inner world: though this may sound portentous, it does not look so when one sees it simply and clearly in the work of very young or less able children. It is just what the children I wrote about in *English for the Rejected* were doing.

Dr Segal gives us another important clue. The capacity to use words poetically to explore inward experience is a comparatively late achievement, in terms of the first stages of the growth of consciousness. It belongs to the 'depressive position' or 'stage of concern' which is the stage at which the infant becomes aware of others, concerned for the effect of his phantasies of hate on them, and impelled to love and 'give' in reparation. However, although the capacity for symbolism emerges at this second stage, it becomes thereafter an instrument which enables us to work creatively on earlier problems, deriving from the schizoid position—a stage before the identity, and the capacity for relationship, were formed at all (and so before the capacity for symbolism *could* exist). So, once the achievement of the capacity to employ the poetic function comes to us, we already have an enormous backlog of symbolic organisation to work at, towards the discovery, and strengthening, of the identity. These inner problems with which symbolism enables us to deal by 'psychic creativity' or 'working on our inward world' are, of course, never solved, since we never can fully understand ourselves (and life would lose all its interest, beauty and promise if we could). So since the dynamic relationship between this inward activity —this quest for identity and touch with the outer world—is never complete or 'solved', we need continually to seek to build bridges between subjective and objective, by symbolism. Language, and especially creative language, is a major medium for this bridge-building, and the consequent strengthening of identity and the capacity to deal with the world.

Where the English teacher is concerned, though he may find such deeper explanations of the function of language interesting—and find

---

* 'Notes on Symbol Formation', *International Journal of Psychoanalysis*, xxxviii, 391.

in them confirmation of the value of his work—he does not have to seek to apply such theories self-consciously or intellectually. He need only exercise the natural impulse to find pleasure in the imaginative uses of words. And his essential concern will be with meaning— 'meaning' here meaning poetic or 'whole' meaning, within the given complex of the symbolism. He does not even have to interpret the symbols: all he has to do is to take the 'given' phantasy, as we shall see with *The Ancient Mariner*. He does not even have to ask of what the albatross is a symbol, in explicit terms.

The value words have in such intrapsychic processes as we outlined above depend upon our ability to respond to them, possess all their richness, and use them creatively. A teacher needs, besides his own relish for language, to be able to share these activities with his pupils. In sharing his words, he is sharing his creative symbolic exploration of whole experience. So, any discussion of English eventually comes down to a concern with meaning, in relation to living processes.*

Thus, despite what may seem formidable if one analyses all its complexities, the essence of it all, happily, is that a teacher of English needs to be able to read superbly, to be able to make his experience of reading articulate in clear terms, to know children, to be responsive to their imaginative life, and to know how and when to make use of literature with them. These things all apply as much to a student teacher who is doing a 'general English' course, as well as English specialists —and even to those who are not likely to teach English, since literacy is the basis of all effective dealings with reality, and in any case 'every teacher is a teacher of English' (George Sampson)—a truism to which lip service is paid everywhere, but which is not yet fully accepted (as the paucity of in-service English training indicates).

Obviously some colleges of education do give such a training: one states its syllabus plainly and simply in such terms as this:

General Course   *English*

Students are given practice in reading aloud, verse-speaking and oral narration and exposition. Some oral work is recorded for the purposes of

---

* It should perhaps be suggested here that, while the new linguistics is valuable in that it takes a fresh look at the structure of language and its social modes, its disciplines are not those of such whole aspects of symbolism and communication with oneself as are indicated here. Because of these limitations, it can hardly be accepted as the dominant discipline in English. It seems unlikely, however, that linguists will accept this, for the new preoccupation is obviously exerting its own persuasive fascination, sometimes at the expense of 'whole meaning', in an attempt to dominate syllabus reform. Some linguists still believe one learns to write by learning the rules first.

25

self-criticism and the discussion and improvement of speech and presentation. Instruction is given in mime, improvisation and methods of dramatisation...

Principal Subject

The purpose of the course is threefold:

(*a*) To enable the student to develop as a person and as a teacher by the study of literature that illumines the variety of human personality and experience.

(*b*) To give her a criterion of excellence in literature by which she may develop her own taste in reading and that of the children she teaches.

(*c*) To afford her some experience of the special disciplines of learning involved in an individual exploration of a topic in literature.

This makes sense, and allows a flexible approach. One wonders sometimes what goes on elsewhere, where the printed syllabus is larded with phrases seemingly designed to impress the ignorant—by laborious reference to 'the function of language as an organic means of communication', or 'the comprehension and appreciation of verse or prose'. At another college the works to be studied include:

Sir Thomas Browne, *Hydriotaphia*; Fuller, Selections; Earle's *Microcosmographie*; Walton's *Lives*; Burke's *A Philosophical Enquiry into the Origin of Our Ideas on the Sublime and Beautiful*; Browning; Tennyson; Meredith; Carlyle; Arnold; and Ruskin.

At a third:

Beowulf; *Le Morte d'Arthur*, Books xiii–xxi; *Redgauntlet*; together with 'one or more of the plays of Christopher Fry'... *Waiting for Godot*...C. P. Snow; C. S. Lewis...

At a fourth the line runs:

Wells; Galsworthy; Huxley; Waugh; Angus Wilson...

At a fifth:

Doughty; Macaulay; Pater...

With one or two possible exceptions, the works and authors in these latter syllabuses seem as remote as could be from the needs of young men and women who are to become teachers. They are utterly remote from the needs of either child or adult in the teaching situation. Many of them have little or no value in the development of any adult's response to imaginative literature, and are of little interest to anyone outside a limited and rather dusty area of textual scholarship, or Sunday newspaper coffee-table talk. Where things have gone this way, obviously some works have found their way into the syllabus through the taste of the lecturers for ephemeral middlebrow fashions, or because the weeklies have pronounced them 'masterpieces'. Besides the question of the quality of attitudes to human nature (which would make me question the inclusion of some of the writers), there simply is not that amount of time to waste: needs are too urgent, and there is too much to be done, even if we only consider the minimum.

The minimum, as I hope to show, must include possession of a number of significant major works in English literature—real possession of their meaning. So long as this is kept in view, it can then be decided whether there is time for attention to some of the academic or ephemeral modern works which appear in the syllabuses. Can there really be time for Wyatt's *Anglo-Saxon Reader* (a college in the north), *Pincher Martin* (London), John Osborne (Southern England)? Even Brecht's *Mother Courage* (south coast)? A student (East Anglia) once wrote to me, 'The books I have read this year have been *The Catcher in the Rye*, *Lord of the Flies*, *Tender is the Night* and *Father and Son*.' She had never heard of *The Secret Agent* and her 'personal prejudice against Dickens' had prevented her from getting beyond the first few pages of *Dombey and Son*. Were the priorities in her training right?

Elsewhere one finds English given over to the kind of novel pronounced nowadays on the book pages of *The Guardian* or *The Listener* 'brilliantly dirty' or 'enjoyably smutty'.Where it is most confident this 'new modernism' in English is often inept. One student at an 'enlightened' college outlined his course thus (after he had left):

First Year: Creative writing (mostly based on Dylan Thomas and Ted Hughes). Tutorials consisting of spelling tests and long boring periods of being talked to about Aldous Huxley. Lectures on the origin and growth of language revealing more about the lecturer's prejudices than about the subject.
Second Year: More tutorials with spelling and talks on H. G. Wells (about a hundred hours over the two years just spent on Wells, Huxley and Lawrence!)*...Lectures on 'modern' novels (Wain, Sillitoe, Murdoch)...

Such an English course, given over to 'modernity', is hollow at the centre—it provides nothing for the young teacher's essential needs (except for what he may have got from Lawrence). And, obviously,

---

* What is indicative is that a man who (apparently) finds Lawrence important should consider Wells and Huxley of equal interest.

there was little close reading and seminar discussion: students were simply lectured to about works which interested the tutor, or were felt to be fashionable.

To sum up my experience of teachers' refresher courses, as external examiner, in my seminar work with students in training colleges, and with graduates in the education year, I find that many teachers have not been trained in the essential disciplines. The teacher going out into the world often cannot read well enough. He finds it hard to take poetic meaning. Students often do not know children well enough, and they do not know literature well enough at first hand. They find the discussion of children's own poems, or simple poems from adult poetry, extremely difficult—if, that is, they are asked to discuss it in their own terms, in relation to their own experience, and not talk the nonsense they have been trained to prepare for 'appreciations' in examinations. Their capacity to refer to the body of literature, with any sense of direction and from personal 'possession', and to relate works to children's needs as they arise in the classroom, is not developed (though they find this exciting as a prospect). Of course, once their energies are taxed by the daily fatigue of the classroom they will be able to do these things even less adequately: no wonder so many settle for textbook work, and virtually give up anything that can be called 'English' at all.

Yet while these essential disciplines are neglected student teachers are spending weary hours at lectures on 'psychology', 'Education', 'aims and methods', and 'Eng. Lit', taking useless notes, which then put aside until the time comes to be examined on them. They they simply practise their training in flannelling examiners.* Even the assumption behind such work is wrong—that by 'knowing' a few facts, theories and precepts intellectually, a teacher can direct his work in a living context subsequently, by conscious application of the tenets of 'reason'. The assumptions behind such instruction are that the relationship between 'knowledge' and living is direct and simple, whereas there is only a confused and complex relationship as I have suggested between people's intellectual knowledge and their inward dynamics. A girl in her first year as a teacher put this point about the difference between theory and life well in a private report:

* Students tell me that in taking their examinations in the Education Year they tend to regress —to the habits they employed to pass O or A level. They do not write in the exam so much what they have learnt but 'make it up' on the basis of the essay-writing tactics they learnt for G.C.E.

When confronted with a difficult child in the middle of a lesson there's no time to start being philosophical about the situation. One just has to *do* something before the whole class get out of control. Large classes tend to aggravate this kind of problem. Having a great deal of preparation, book marking, and generally keeping one step ahead (or at least trying to) does not leave much time or energy (and certainly not the inclination!) for calm, rational thinking. The problems are of such a different nature from the frequently purely intellectual problems of college work. Thus it is much harder to come up with a practical solution that is going to be effective. I think that one of the things I am having to learn at the moment is to have a diagnostic approach towards the children's reactions, and then, having attempted to pin-point the trouble, to think up what needs to be done to help the situation, and finally to translate this into terms which will be effective in the classroom.

This sort of approach is completely impossible when one first starts teaching. For myself I thought I had worked out what my approach would be but when it came to it I found that my reactions were very often haphazard, inconsistent, and certainly not very rational...

Of course, some of her discoveries could only be made through experience. But it is her education which to some extent misled her, in encouraging her to suppose that in her relationship with children her reactions *could* ever be 'rational' or even 'consistent'. Or that situations in life—whether in teaching or not—could ever be met by 'calm, rational thinking', or by 'being philosophical'. A classroom situation, like a critical family situation, comes suddenly upon one, and one has to live through it, from hand to mouth, relying on one's intuitive powers (as a mother does with a baby). This is not what many lectures on the theories of education and psychology imply.

On the other hand, time spent reading *Women in Love*, or Blake's poems of hate and love, or George Eliot on Maggie Tulliver, or *Huckleberry Finn, would* convey in terms of 'felt' meaning the nature of that turbulent feeling between adult and child that tends to emerge in a classroom situation. Even an honest (if not creatively exceptional) novel about the experience of teaching, such as *To Sir with Love* or Blishen's *Roaring Boys*, will convey more of the 'felt life' of the work, often by incidental asides and by giving the 'feel' of episodes, than a dozen books on aims, theory or method.

So one considerable change which seems to me urgently needed is for the student teacher to have much more experience, early in his course, of teaching children—and then afterwards, instead of lectures, to have real seminars, to discuss the nature of children (and of Man) from the experience of the classroom, and from imaginative literature afterwards—with some urgent sense (chastened by the hurly-burly) of the relevance of these discussions. Every young teacher writing to me about his or her experience of training urged more teaching practice, and *earlier* teaching practice. If we accept that teaching is an art, relying on intuition, and being made more awake to metaphor and aware of meaning, then teacher training must start with 'whole experience', and go on as a process of the exchange of explorations by the conscious intelligence, between adult and young person—around actual children and their work. This would be to make it an experience of education as exploration, not as the passive 'taking in of information'.

This book then seeks to define what happens between adult and child —and how they meet in the word, and how they may develop a more acute attention to meaning. This will inevitably involve a consideration of what literature is valuable for the young adult, and the child—what, from the body of finer expression, can be used to foster and illumine their own work.

So, at the end of one's exploration, one comes back to insisting that our capacity to educate comes by nature, and that even Dogberry was right in a sense in saying that 'reading and writing come by nature' too.

NOTES FROM
THE UNDERWORLD

3

AN EVENING IN SIR'S
BEDSITTER

Let us imagine ourselves perched on the back of the chair of any young
beginner in the profession, sitting one evening in his digs in a back
street in Walthamstow or Salford, with a pile of dog-eared exercise
books covered with blue sugar-paper, inky, and smelling of chalk
dust. A faint echo of the buzz of eight noisy periods of the day still
rings in his head, and his hands still seem to smell of the fishy food
containers he helped to clear away at school dinner. All high-sounding
pronouncements on educational aims, all the psychology lectures about
child development, come in the end to this: thousands of weary teachers
every night trying to rouse sufficient energy to thumb through children's exercise books and to mark them, while trying to estimate what, if
anything, they have achieved.

With a sigh our young initiate, who we may suppose to be teaching
in a secondary modern school, notices that even the covers of the books
are scrawled with indecencies: how can poetry live in this atmosphere?
On George's book is the legend 'Julia is an old shag bag', added to the
Road Safety rules:

ROAD SAFETY

1 Never run straight out on to the road when you leave
   the school premises, or when you leave your home.

2 Before you cross the road look to see that no traffic is
   coming from any direction.

33

When he is tired his failures will be more apparent than his successes. The mere appearance of children's work will depress him. There will be books whose covers have been garnished with crude war drawings. There will be poems which seem to have been idly scribbled. Do they mark the failure of a lesson? What should he do about rude words scrawled in the margin by the unbalanced pupil? Is there anything here to commend?

6-12-60	CHRISTAB POEM
.	When its christmas it something different.
.	But it soons goes —
. . —	The Ground is almost solid,
. . .	unless it beggins to snow,.
— . .	but not very likely this year
b . .	it will not be rain like hell.
.	The turkeys a begin to gobble
— . .	but there nects will be strecked
— . .	they will be stuffed with sage & onion
— . .	and washed down with a dron of gin
— . ✓	You will be sick if you have to much

In a girl's book he will find two utterly different kinds of handwriting by the same hand: neatly written fragments from Enid Blyton, alongside violent scrawled imaginings from the world of pulp-books. Child and adolescent appear day by day in the same book like split aspects of the same pupil.

kidnappers
Eg Every 16 hours
the girls were injured
and told them that
they didn't do as they
were told they would
be touched. They
would hang by their

NODDY

One day as noddy was going down the road in his car and as he came to the corner he nearly knocked over MR. Plod the Policeman.

arms so that their feet would not touch the ground.

Things scrawled by children in the bottom streams, with their fragments of flip talk from television plays, will seem at times like the graffiti of madmen.*

when your 21 a your ~~eggfats~~ ~~egg~~ toatch your
~~a history~~ your a man

'its a poor poor world
with poor poor people
with poor poor fags
and poor poor wine
with poor poor women
and disgusting young men

The agonising scream oj the tint
made your blood turn cold

* See *The Secret Places*, p. 25.

Attempts at creative writing will produce from some boys only echoes of the worst kind of pulp war story:

heads down. The nips charged again and Joane gave them short blasts with his bren when Tony Smith lifted his heavy 303 and began knocking them down one after the other, but with drums beating and trumpets blowing and insane animals cries coming from their throats the japs advanced, the men went down under the hail of concentrated fire at the end only Jones was left the nips got him alive and they staked him out on an ant hill and so the last of the platoon died screaming out in angony while the nips looked on in delight

The End

Sometimes even the handwriting is distorted to a degree that seems pathological:

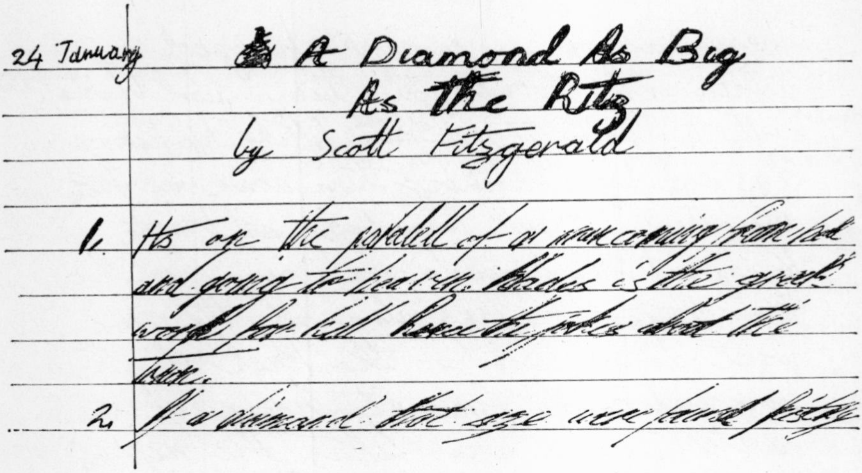
24 January — A Diamond As Big As the Ritz by Scott Fitzgerald

1. [illegible]

2. If a diamond that size were found [illegible]

Sometimes it is the content:

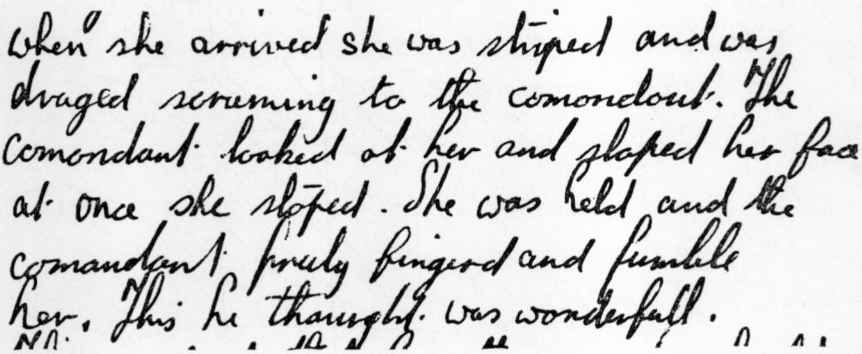
When she arrived she was striped and was draged screming to the comondont. The Comondant looked at her and slaped her face at once she sloped. She was held and the comandant freely fingerd and fumble her. This he thaught was wonderfull.

Some pieces he will not even be able to read:

The ghost of Coventry

Owest · upon a time there was a
man is nim was Mitor witor and nedad
wonon was Keid and Mitor witor was
a couast and ne was Gilve and to
hab By the nec bill beide Be for
He was browing to Be hab He sdis to
three man I will kill Boy miton witon
side They The Gog son is one one
ow Keid the wonon annd He wet
to a Lhoc to met a · man to pags
ffon wat it Bon then the Boys
Kted the man and ran home to the
Gog then the Boy foo eat d Bot
the kilin of to wonon dad the How
He for out the ton of the boys
fill out of is Pobit and son Bad
to the Boy I will kill you and
Goy Bon the pes and the mon kina
is firthe and the Bor sedo way Pis
you killn Gor saythe sedo is Docor
The END

He may have set out to pursue this work instead of the work he was supposed to follow by the official syllabus: can he justify it? Has he any allies in the staff-room?

Almost every headmaster who came to lecture to us at college about some aspect of education assured us the future was 'ours'. Some of us are still clinging to the shreds of those early ideals, but many have already given up the fight and succumbed to the enervating cynicism that deadens so many staff rooms. 'When I was your age I had ideals—you soon get over it', they tell us. 'Oh, I get by. I'm not sticking my neck out...' 'I don't believe in this system, but I'm not doing anything about it!'*

From time to time there may be a triumph. There is a loose piece of paper: Jack's brother has typed out a poem for him:

```
Ans some dark night when the moon is bright, and death
stalks on the dale; I see that room and I know real fear
real fear; My hand shake and my musles tighten and the
        veils
        ──── of time just drift away; And horror gasps in quiet
dismay And the blood drips like a brilliant dye and stains
the floor a brilliant red, and their is no-One to speak to me
of life, love and liberity and hate drops in where once
was love. And i seach in that room for one speck one speck
of hope to light the doom But none comes and none ever will
and i'll live in this enternal hell where no man lives,
                                        lays
where no man loves, where no man ────, where no man dies.
                            and
But where I'll spend enternatiy, I think of things that used
to be, Of days gone by of memeriors that still ponder on
        But
my lips. ─── bloodstill runs on the floor and death still
knocks on the door. And I wonder if in death I'll find
peace like I had once before  IF YOU like Dont like
```

on that cold dark and dismall moon a long long time ago. the E N D.

It is startlingly better than his first poem, written on a loose leaf in his exercise book: this is the kind of achievement only apparent to a teacher who knows his pupils.

But, even though something has 'panned out', consider the effort and the skill required in assaying it! In this poem alone we have material for some fifteen minutes' study: and to make such a study a teacher needs to be able to read well. Remember that he may have ninety books to go through this evening. He needs to have enough of a literary

* These quotations are from reports on their first work as teachers by students from various centres.

'background' to know whether Jack is reproducing poems from an anthology, and then to judge the value of the boy's peculiar brand of melodramatic utterance. To find the preoccupation here with love and hate Blake-like ('and hate drops in where once was love') and a unique and personal expression of adolescent fears of being annihilated by conflicts of black feeling requires a sensitive response to poetry. If he has this the teacher will rejoice in such verbal felicities, naïvely original as they are—'the veils of time', 'ponder on my lips', the 'felt' language of 'my musles tighten'. He will be able to admire the groping rhythm of 'And i seach in that room for one speck one speck of hope . . . But none comes and none ever will and i'll live in this enternal hell where no man lives, where no man loves, where no man lays, where no man dies . . .' These beauties of verbal expression, and the poignancy of the recollection of untroubled childhood, during the 'schizoid episodes' of adolescence—'peace like I had once before'—are far more important to note than to fuss about the misspellings 'memeriors', 'liberity' and 'enternatiy': the important positive thing is that Jack is not afraid to use such words, even though he cannot spell them. Now, the teacher must consider, what do I tell Jack? What do I do with his remarkable poems? What, from literature, can I use to lead on to, from this gift? (*The Ballad of Reading Gaol*, by Oscar Wilde? Blake's *Poison Tree*? something from Tennyson's *In Memoriam*?)

The examples I am giving in this chapter follow in no 'graded' or progressive order. They come as they come in the teacher's life—at random. Is this an adequate response to an exercise in 'free writing', in response to a gramophone record of Varèse's *Ionisation*?

Ionization

This piece of music sounded like big bomb explosions in an atomic war, with gusts and when those sirens go it makes you think of waves of bombers coming over while the sirens warn of their approach, certain clatters sound like atomic machine guns and big bangs sound like atomic bombs, big screeches sound like huge buildings being blown up and thrown to the ground. I thought that this piece of music put a sense of action and excitement into you.

Of course there will be much that is uninspired. But the reader will see that our young teacher is involved in something deeply satisfying and sensitive, as he sits by his gas-fire. Yet he may well sigh over the difficulty of fostering such sensitive matters in the atmosphere of school to which he is to return in the morning:

At Assembly...everyone sat in a sort of numb silence for the Head to sweep in, a hymn was rushed through by a few voices, an inaudible reading was given, prayers of high and noble intent were said by the Head, the Lord's Prayer was mumbled, notices were read, and everyone was shunted out again with cries of 'Keep to the left!'.

Even if he is human, can he count on his colleagues to be?

The staff at my first school were pretty much agreed on what the ideal school product should be. By far the most important quality was presentability. If a child was 'well turned out' (i.e. school uniform recently washed and well-pressed) all failings in character and ability were o'ershadowed. ('I'm glad Mary has gone down. She doesn't *dress* like an A stream child...') 'Intelligence' consisted of ability to assimilate and retain the knowledge and petty skills necessary for G.C.E., U.E.I., R.S.A., College of Preceptors, and the various internal examinations that led up to these pinnacles of learning. A boy with outstanding drawing ability was not allowed even to apply for entrance to local art school because he had been caught masturbating in the boys' lavatory a year previously.

So, much more is needed than the capacity to read well *and* quickly: he needs to be as mature and self-confident as possible as a person. Little boys are often crude. While yet still children, they also want to try to be bold and much of this impulse is innocent enough: how do we deal with a story like this in class?

Falling In Love,

Dick was waiting outside the cafe for Julie. He could hear the sound of the juke box in the distance, and the clanging of cups. He saw Julie coming towards the cafe. She had dark hair, and blue eyes, she was very attractive. She wore alot of make up. Dick was tall, and he had dark hair. He wore pointed shoes, blue jeans, and an Italian styled jacket, blue shirt, and a dark tie. He had sideboards that came halfway down his cheeks. Julie came up to Dick, she was very sexy. Dick said, "Shall we go in the cafe?

Julie said, "No it is too stuffy in

there, let us go into the park, it is a nice moonlight night." So Dick put his arm round Julie, and put her arm Julie Dick, and they walked through the park. They sat down on a seat, which was under a tree near a pond. . She had lovely breasts, and a good figer. He said, "Julie I love you very very much."

Julie said, "I will always love you." There they were siting on the seat, kissing and cuddling. He put his hand behind her back, a started to undo her brassies, She t slapt him on the face veal hard which made him jump back. She said, "I hate you for that." Dick was very angry, and said, "I hate your liver." She marched off in a vage. Dick was very sad, and he went back to the park were he had parked his Norton. He was mad, started his motor bid bike, + pulled away very fast. He was travelling down the high street at 8×5 rals m.p.h. he

44

must of gone mad, he took a blind corner at 60 m.p.h. crash, he hit a pig lorry, with a head on colishon. He was kill not instently, blood trickled down the road from a gash, in his head. Julie heard that Dick had been killed that night, she burst into tears, because she thought it was her fault, she could not sleep that night. He had gone, but her love for him still remained.

Of course, a student teacher might protest hotly that he was not going to be the sort to allow children to write for him in such a shocking way. But every teacher gets something like this from time to time, in one way or another. Peter is really frightened of sexual experience: this is what lies behind all his cheek. So, symbolically, in remorse and retribution, the hero is smashed to pieces at '60 m.p.h.'. The death-by-the-motor-cycle is perhaps a symbol of fear of adult prowess, and unconscious fear of sex (see *The Secret Places*, *passim*). Yet the child tries to cover his fear by being as bold as he can, as little boys do, taking things from television plays ('she had lovely breasts, and a good figer') and 'big brother' talk ('started to undo her brassiurs'). The violence consequent on the love-making, love turning to violence, however, seems oddly real. (My own reaction would be (*a*) to say nothing which could be interpreted as praise or blame—the story is not really very interesting as stories go; (*b*) not to read the story out, and to suppress any disorder in class such as Michael might cause by showing it to others, tittering, and so forth.)

But every teacher needs to work out his reactions to such awkward moments: and how he does so is essentially a literary problem. A student teacher needs to be given an indication that some little boys will be obsessed with sex, and may seek to make up for their miserable feelings of puniness as they sit beside girls who are so much more advanced physically than they. So they will put such rude things in their stories, or swear horrible or write on lavatory walls. At root the problem is one of trying to boost a weak identity by appearing to be big and bad. The English teacher's defence where he comes across such 'rudeness' in a story is to maintain a concern with the story as English, in discussing whether it is interesting or not. Above all, insights gained from literature will help a teacher avoid moralising, and especially rejecting.

Even average daily work continually throws up such complex questions of writing and its quality. What I am trying to convey is the need for a teacher to be able, in class, or with a pupil, to make rapid and sound 'literary critical' comments, being both severely critical to himself while being positive to the child, *especially* over rather mundane efforts. From this work, too, she should be thinking how the child may be stretched by literature from the shelves. Peter might write a genuine exploration of courtship if he has read some short stories by Hemingway about Nick and Margery, or has listened to some folksongs, before his next attempt.

The teacher who works in this way is able to foster a natural interest in what literature can offer. But only from a very good literary training of his own can he have confidence in knowing the right thing when he is 'given' it.

The most remarkable poem I have ever been 'given' by a child is this (it is discussed in *The Secret Places*, p. 21), by 'Robert' in 3B. It would seem to me a classic for student teacher discussion, being so obviously an expression of the Oedipus myth. It is besides the best example I have ever come across of the creative surprise—a piece of work one could never believe it possible to receive from such a tough, alienated, and apparently sophisticated boy of fourteen. Without a sound knowledge of the more gnomic English poetry a teacher might miss such an achievement altogether: it does not look much, after all! Would our new young teacher see anything in it?

One day a man killed me.
I have been looking for him.
All my life.

I am a ghost how thee me.
I found him with my mummey.

I killed that man
like he killed me

he die on my ~~~~ .
Mummey ~~ nee .

But even an obviously good piece of writing may challenge the ethos of the whole school, and so the teacher's security in his subject will be challenged.

For instance, a student teacher obtained the following piece of writing from a 'difficult' girl of fifteen in a grammar school. It seems to me a remarkable expression of an adolescent's feelings of isolation, of a lack of feeling, of loneliness, in the crowd of adolescents seeking distraction. It communicates the sad anguish of a child who cannot break through the barriers and find relationship—and the spiritual chilliness of being 'shut up' in oneself.

The student teacher showed this piece of writing to her headmistress, because it was something of a 'break through' to a 'difficult' pupil. The headmistress said, 'Yes, that is very good—but it is a pity, isn't it, that it was ever written?' That is, for her it came too disturbingly direct from the world of coffee bars, motor-cycles, and Mods and Rockers—

and so was alien to the respectable grammar school ethos. Humanly speaking it is a marvellous achievement, for this girl might well have found, in communicating thus to the young student teacher, a way out of her isolation. But to the grammar school head such a breakthrough is something of a threat, because it speaks of an area of the experience of the adolescence by which her standards are challenged and the authority of her experience doubted.

Later, when the same girl was arrested for shoplifting the student teacher said she felt she was being made to feel she was even responsible for this lapse in some way—by indulging the girl's self-expression!

Saturday nights

I sit in the coffee bar watching the dim, orangy-yellow lights gleaming thru' the haze of cigarette smoke. They remind me of something but I am not sure what. The atmosphere is thick and heavy, and I feel lazy, and withdrawn from everybody around me. I can hear the babble of conversation without hearing the words. I do not want to hear them. I can see a blur of faces in a dream-like way, and I sit half-asleep thinking of nothing in particular. Suddenly a piercing voice demands 'Will you take your parker's off please!' I come back to the present and see the boss-eyed waitress standing in front of me. She turns on her heel and walks off and I relapse into a daze. I am watching the yellow light at the end. It is barely visible through the smoky atmosphere. 'Take your parker's off *please.*' This time with more emphasis. There is a sudden scuffle of chairs and everyone stands up. I cannot be bothered to go out into the cold. Before I know what has happened I am downstairs. I shiver with cold, everyone is cold, the scooters start up, I am on the back, and now we set off.

The scooter wobbles slightly from side to side. I am cold. I turn round and see lights everywhere behind me. The scooters are all ready to go. I am cold. There is a sudden outburst of noise. I am frozen. Then with a roar of engines we are off. We are in the middle. We overtake the scooter in front. we are overtaking the next, and the next. We are in front. I can feel the wind has blown all my hair back. My eyes are watering. The blood is pounding in my head. My eyes are streaming. My ears are getting cold. I put my head down and the wind rushes over my head. A scooter has come up and overtaken us. We go faster and come up beside it, we are side by side, we are in front, I feel exhilerated, I feel like shouting, I feel happy, I wonder vaguely if my ears will drop off with the cold. My legs are numb. We are well in front. I think we are doing about sixty-five miles per hour. I wonder if my eye-liner has run? I expect so. I do not care. I put my hands in my pockets and lean back. We are travelling along steadily. Suddenly we slow down, the rest catch us up, the scooters quiver impatiently and the engines roar noisily. Silence. It has become quiet. The engine have stopped. We have arrived. I am getting off the back. I am standing on the pavement. All of a sudden I feel very tired, and cold.

Such an experience reveals that good English teaching may bring a clash with authority because it evokes human energies which colleagues cannot deal with—because they have decided to cope with life and teaching by ignoring large areas of experience. What shall go? The blindness? Or the creative challenge? Of course, the change to comprehensive education may itself, rather shatteringly for some, cure the blindness at least.

Such writing inevitably raises questions of the child's personal life outside the classroom, beyond 'mere' English teaching—though the *words* remain the focus of the teacher's concern: how 'ill' is 'Robert'— can he face life without help? Why does Susan express such deep isolation? Is it connected with her theft? In seminars I find students doubting whether such considerations are really within the English teacher's province. Experience in the secondary modern classroom compels one to accept that it is impossible to avoid such concern with living needs outside school: it would be a betrayal of one's profession *not* to hear such human cries of distress, or to shrink from risking such creative exploration. The point is made clearly in E. R. Braithwaite's account of teaching in a London school, *To Sir with Love*, and also in a different way in the book on school phobia, *Unwillingly to School*, by Kahn and Jean Nursten (Unwin Paperbacks). A child's learning problems often lead one to need to consider the whole family complex on which, as a child, he is still dependent. But, of course, the English teacher's essential function is to foster a child's capacity to deal with all experience through language as well as he can. Yet, obviously, what teacher, reading such a piece as the following, would not feel, as a human being, that he ought to keep life-lines out, and even to discuss the situation with his superiors?

An eventful night

A few weeks ago I went to stay at my Auntie Joyce's for a day or so. The first night I could not sleep at all, my Auntie and Uncle were shouting and arguing downstairs. I heard my Auntie shout me and I thought, 'Good Heavens! He's killing her', and I said to myself, 'now be brave Jill, go and stop him.' So I jumped up out of bed and ran downstairs. My auntie had a knife in her hand and My Uncle was behind her holding her arm. Auntie Joyce slashed out with the knife, I just stood there shaking. My Uncle said,

'Will you take the knife off her or I will have no ear on.' So I leaned forward and took the knife slowly out of her hand. Then Auntie said 'Tell him to get off me,' so I said 'Uncle, just get off her you mad, raving lunatic.' So he did and turning to me he said 'Are you threatening me?' I said 'No, Uncle, honest'. My Auntie ran indoors and said 'Get back to Scotland,' to which he replied 'Joyce dear, calm yourself down', which got her mad so she threw all his clothes out into the street. Then she threw a bottle at him. It hit his head, bounced off and smashed all over the road. Then he really was mad and he seized my Auntie. She shouted 'Go for the police,' but I went round the back, crept up behind him and brought the poker down on his head. He fell to the floor cross-eyed. My Auntie said 'You've killed him.' I fetched some water and threw it over his face and as he came round he said in a stutter 'What happened?' I could not help laughing as we took him back into the house. By the time we got to bed it was ten past five in the morning. The next morning when everything had been straightened out my Uncle said 'Joyce you had better tame your temper, you'll be killing me yet,' then he looked at me and said 'Well at least your niece sticks up for you.' After dinner I went home and told my father about it and he said 'You won't go there again, your Auntie is a nutter.' Oh! I never did find out why they were arguing.   (*A girl in a secondary modern school, aged 15*)

Of course, only by experience does one come to be able to know the degree to which a child exaggerates and embroiders (as I think she does here, somewhat). But to teach English properly is to accept that one is dealing with human beings, and so one is obliged to be human, too.

So, if we are human, we can perhaps tolerate the adolescent's need for the exploration of the sensuous life, and the need to test the teacher by what she writes. To be prepared for these things beforehand will help a student teacher deal with such pieces as that opposite, as they come from the writings of bottom stream, and middle stream, third- and fourth-year children.

It is a sad and pathetic piece with childish moments of bravado that would horrify the grammar school headmistress mentioned above. Yet any teacher seeking to get through to adolescents in the secondary modern or comprehensive school needs to be trained to face the problems it raises. For his pupils not only have their own compulsive anxieties which impel boldness: they also have a diet of smut and grossness thrust at them from pulp-books, cinema and television, much of it very destructive. It is impossible for us to close our eyes to the effects: and here, after all, the preoccupation with marriage is plangently childish enough. For me this scrawl is made all the sadder because I

A date in a lane

One day I was walking down
the street and as she went
down the lane she met a man
and is name is Peter Hicks and her
name is Susan Eaton and he made
love to her and she said Peter Stop
it you are not to do that my mum
well come and see you in a Tick
and then you came ~~strikethrough~~
with your hand on my back and then
you can make love and then
they got engaged, he buy a ring
of gold and her mother was
so glad and then they want
to the pictures and he said

well you marry me and she
said it to soon to be married
yet, all then we well get married
one April the 21st is it all right
to get married then so she
said all right it is all right
to marry him then so I can
marry him on April 21st yes my
dear yes my dear you can. I am
glad that you are glad; so
she married him in the End.

met the writer (who was illegitimate) two years afterwards, with a new-born baby on her arm, which she obviously loved desperately. The following conversation took place:

ME: I bet you didn't know you could do that!
JANICE: I had a good idea.
ME: Did you marry someone from the school?
JANICE: No. I married a Scotsman, six months ago. We're separated.
ME: That's a pity isn't it?
JANICE: No. He was a swine!

At eighteen! The time to prepare children for life is short enough! So, a *cri de cœur* such as the following cannot be ignored:* receiving it made a great difference in the relationship between this boy and myself:

> Come on, and Sigh with me
> I have a mind which is blooms confused
> I am such, so sick in the mind.
> So hear who I have been used,

Such work from children, which raises difficult questions of personality and relationship, is, of course, in a minority, though as one goes down to the 'lower' streams the proportion of such disturbed work increases. It has to be seen in relationship to the large amount of excellent 'normal' work by stable children of middling ability. There is much that is merely straightforward and competent— an account, say, of a fishing expedition by a boy with nothing 'symbolic' or 'deep' or 'disturbed' about it. The teacher recognises this as the work of a healthy boy writing simple good English about an activity in which he is greatly interested (see example on p. 53).

But the reader will by now have taken in the effort needed, even to read one small boy's painfully written piece, to relate it to his achievement and his character, and to make notes on following it up. And to face the bigger problems! The teacher may have 200 books to mark in the week! In considering time I am concerned merely with *content*—I am assuming an agreement that there is no call to mark every literal fault (though there are schools where every tiny mistake has to be

* See *The Secret Places*, p. 25.

marked). Even to read, take in and understand sixty such pieces in an evening (the work of two forms in one or two periods only) is itself an exacting task enough, leaving aside the questions of grammar and spelling which inevitably arise. The main problems are those such as literature raises, are they not?

Carp

The time was 5 am in the morning it promised to be a good day. I dressed and had some sandwiches for breakfast. Today I was going to Hilton mere pond to catch (if I could) some Carp The bait I had made the night before was ready to be packed. I checked to see if all my tackle was in the bag it was so off on my Bicycle I went.

At halfpast five I reached Hilton mere Pond I tackled up 10 yds from the Bank I decided to use paste and put a piece the size of a shilling on the hook. I made a cast some fifteen yards near some weeds, then I sat down and waited for a bite. Carp are the most suspicious fish and half an hour later I saw a good one come up to my paste which was floating on the surface.

Now, after this experience of a bit of 'marking', how does the reader feel? How is he bearing up? We are only doing, under optimum conditions, what most teachers of English are doing many working nights—certainly those who teach the two-thirds of the school population which goes to the secondary modern school (or through the less academic streams of a comprehensive one).

Let us pause here, then, as if we were that weary teacher, making himself a cup of coffee after looking through his piles of books. Out of our piles of books, in an hour or so, we have discovered a number of complex problems, about our relationship with our pupils. We dimly recognise that these are bound up with 'English' work. Our eyes are weary with reading their scrawl, and our capacities to make rapid assessments of their work have been taxed in the extreme. It is now nearly midnight, yet before next day we have to make decisions as to how to deal with these pupils, in the particular atmosphere of our school community. We have to prepare ourselves to deal with all kinds of children, and to bring them into touch with English literature according to their needs and interests.

Perhaps there is very little in our day's harvest of books that seems to justify our allegiance to the 'creative' approach to English. Perhaps during the day our beginner has already had a battle with the head of the English Department, or the headmaster. How much inward conviction has he, that he is on the right lines?

How much of our beginner's college course in English has helped him to deal with the day's work, with his study of children's work—and with his preparation for tomorrow? Does he feel, as one first-year teacher writes below, that even a good college training falls far short of what is needed?

Was my training relevant to what I found myself doing in the classroom? The answer here is very largely 'NO'...it was, of necessity, too idealistic. Something which has struck me with more force than anything else since I started teaching is the terrific gap between the ideal educational situation presented at college and the actual situation one finds in schools. My biggest problem has been how to bridge this gap. It is very obvious that it cannot be done quickly but I feel that I don't even know how to *start*. I know very well how I would like to finish up! I have a very clear picture gained from College of how I would like to be able to run my classes, what sort of work I would like to teach, what type of approach I would like to feel I could use with the children, but all this is impossible because of the general atmosphere of the

54

school and the nature of the work that the children are used to. For example, the children I teach actually *like* dull, repetitive, habit-forming work. Anything original or unusual is too much for them and they cannot cope with it. They simply do not understand. Although at college I learnt how to plan work for teaching on a constructive basis which would make the work interesting for the children, I never learnt how to plan on the sort of basis which I have inherited in the classes I now teach. One is up against almost all their previous experience. The problem is how to get them out of this attitude before I can even start thinking about any 'training-college-type' work with them.

Of course a lot depends on the type of school, what kind of children they are and what they have been used to before.

As regards teaching English these are the main problems I have come up against so far:

(1) Lack of knowledge on my part of literature that is really the sort of thing children will enjoy. This sort of thing seems very remote at college but I think more could be done about it during the teaching practice periods.

(2) Lack of good, modern books to use with the children. There can be nothing more harmful, I feel convinced, of presenting a child with a dull book that looks as though it has just come out of the Ark !

(3) Basic problem of lack of vocabulary in most of the children. (My teaching is more or less limited to 11 and 12 year olds.) I am certain that television and other mass media entertainment are to blame for this. I should not imagine that many of the children I teach spend much of their spare time in conversation with adults. Very few of them read to any great extent, and it is very obvious from their work and speech that those who do are the rare exception.

(4) Lack of sensitive response in many children, or else stereotyped responses (e.g. 'something exciting' is always crooks, a chase, guns and Police). Those children who take an *active* part in any English work I present are in the minority. The rest are quite happy to sit back and remain passive. They do not seem to have learnt, or experienced the reward that comes with being involved in something, certainly not emotionally, anyway. They are very down-to-earth and worldly.

There seems to be a need for more basic use of English—simply how to read a book to enjoy it, how to understand forms, how to explain things intelligibly, how to follow instructions clearly, etc. I would not call this sort of work 'sustenance from literature' but it is, in merely a practical sense, helping them to 'engage with life'. It seems an awful pity that we should have to get bogged down in these mundane aspects of English but it really seems impossible to attempt anything else beforehand. The children simply have not got the vocabulary or the experience to respond to it. Yet because of my own limitations and inexperience I feel that instead of being the medium through which the children can broaden their outlook I am in fact imposing further limitations on them as a result of my own narrow capabilities. My teaching technique seems so clumsy that I cannot present to them the wide scope that I feel they ought to have. I find it hard to think up the right sort of ideas, the most suitable approaches, and it is even harder to summon up enough confidence to put them into practice.

A student trained in an exam-bound college, with an academic syllabus, is likely to find the distance between his training and the class-room even greater. He will certainly be unable to see the creative achievement in 'A man killed me' or 'And some dark night'—the kind of art which even 'low stream' children are able to produce.

A set of questions for third-year English specialists at a men's training college arrives at the time of writing:

> 1. 'There can be no such thing as a mature Romantic attitude, because Romanticism cannot accommodate humour.' Discuss.
> 2. 'The anti-heroic is a characteristic confined to 20th century literature.' Do you agree? Make reference to any appropriate works in your answer.

These seem like invitations to shadow-boxing, round such remote subjects as the Romantic Revival, or *Waiting for Godot*, while the inscape of true creativity may not be grasped. Meanwhile I know that the staple of reading in these men remains not very acute or subtle (do they really prefer Conrad and Lawrence to Amis and Ian Fleming? If so, could they say why?). Such academic generalisations are light-years too far from the work the teacher will be doing—though this is *not* to say that the capacity to read and discuss (say) Wordsworth's *Margaret* or Joyce's *The Dead* would be—for both would be appropriate to the teacher's consideration of *Saturday Nights*. But such exam questions are remote from a human concern with words in literature, in any case.

The problem raised by the handful of fragments of children's writing in this chapter are many. Some are disciplinary, but cannot be isolated from considerations of rejection and tolerance, and the aspect of 'giving' in free writing which is bound up with these issues:

(1) What is to be done about children scrawling on their covers, in the margins, etc.?

(2) What is to be done about obscenities, which seem to indicate that children are testing me out?

(3) When do I read a child's work out and when not?

Others are positive 'leads':

(4) Can I not persuade George to do some full-scale painting of themes from his war stories?

(5) How can I make the best use of the marvellous piece on *Saturday Nights* and of Robert's poem?

Others are prophylactic and therapeutic:

(6) Is the boy with the collapsing handwriting in need of psychiatric help? What is the family set-up?

(7) Study Jack: he is probably just suffering normal adolescent despair and will deal with it himself, but it might be a cry for help (who else could give it?).

(8) Study Robert, as an example of how to break through to an unco-operative teenager.

(9) Have the author of 'Come on and sigh with me' stay behind for a chat: I think we might get on better. But how can I tell him I get the message?

Others are personal: who to compliment, who to give special marks to. Other points involve research—what story can I read them to follow up *Saturday Nights*? *An Outpost of Progress*? Shall I read the class that produced a poem on *Carp* Ted Hughes's poem, *Pike*? or *The Big Two-Hearted River*? What went wrong with my poetry work this term? Is it me, or their adolescence? Dare I read out the piece about Jill's aunt and uncle fighting? Is it 'real' and 'sincere'?

Why did that girl suddenly go from 'Noddy' stuff to that queer scrawled account of girls being whipped? Does she get pornography from the store down the road? (Ought I to write to the N.U.T. for advice on this?) How can I counteract their preoccupation with violence? What does it mean? Will they work through it? As they do—will their English improve?

Here are some of the teacher's deepest problems, and how he tackles them depends upon his capacities as a sensitive *reader*, as a teacher assured of his intuitive powers and as a creative person. How much from psychology or education lectures can help here? Surely much less than much reading from poetry and fiction? If the teacher has been trained as a good reader he will be able to see how much real achievement there is in some of these pieces. The achievements are his pupils', and he can be glad for them, if he is a creative person. But he has been able to prove himself capable of receiving them and dealing with the complexities which arise out of the situation he has allowed to come to life. So, in working with students, what I am always most anxious to get from them is an appreciation of the effort that goes into a child's achievement, and from this to realise that the most valuable thing to say to any pupil is 'well done!'.

If he can see the work in terms of effort and achievements, the teacher can have hope—and he can tolerate the torments and queer by-products of adolescent expression. The violence—even the rudeness and bravado—he will be able to 'take', as a mark of the fear the adolescent feels, at the new feelings of puberty, and of the insecurity of his emerging identity. He can try to hold the ring while the child explores violence, or destructive facetiousness, or just dull patches in himself—hoping that he will come through to calmer and more positive things, to gentler and deeper perceptions, and to more important subjects such as developing relationships. He will recognise too that attempts to involve children in violence or sex too explicitly (in the *Billy the Kid* manner) may come from doubtful motives—he has to keep telling himself that under the blackest jacket is a frightened child. And he will be aware that his work goes on in a world in which a great deal of irresponsible exploitation of the feelings of the young is being exerted by commercial entertainment.

But to cope with all these problems, he needs to know what are the more rewarding themes in children's writing and in literature, and this knowledge he must get from his training. From it he should learn how he can meet the child in the word, and gain a capacity to read and respond to the expression of both man and child, with a sense of relevance. Then he must be able to turn to English literature to find his way about it sensitively and with insight. What kind of training in English gives a teacher such capacities? To answer this we must first look at things as they are.

THE PAPER TYRANT

*A Study of the Content of Traditional English Work for a
syllabus geared to an 'Eng. Lit.' Examination*

4

THE WASTEFULNESS OF
BEATING THE CLOCK

For some years now I have acted as external examiner in English for a number of colleges of education. In the course of this work I study the question papers, and suggest alterations, and I supervise the marking of the papers by the internal examiners. I also study the work during term and hold seminars with the students on special visits. Everything is fairly and efficiently carried out, and I am impressed by the pains taken. Indeed, because of the exaggerated awe in which examinations are held, the elaborate pains can almost seem to give the procedure a meaning that, looked at closely, it can scarcely justify. What the exam becomes is an artificial authority which the staffs of colleges invent and maintain themselves, while almost seeming to pretend that someone else, some objective *alter ego*, is doing it all. By this means they give themselves a false sense of security, and a sense that there is more 'objectivity' in English work and testing than is really possible.

The most one can say for an exam is that it may help relieve the anxiety of working in a subject such as English in which so much is intangible and inward—really inaccessible. An exam seems an outward reassurance that something is 'really' being achieved. It may, however, develop into a paper tyrant which no longer even tests what the teachers wish to be reassured about—their own best work—but other false disciplines which stand in lieu of these. The odd thing is the way this happens without people being aware of it: in several places the staff were incredulous when I said I thought the examination played far too large a part in their assessments and efforts. They protested that 'much more went on' than appeared to the examiner—the examination was a very minor part of the work. Yet I could see that even essays written in term were a training in what an *exam* required, rather than what the subject required. And, in any case, I replied that if the exam was obviously peripheral (and, as I believed, concerned with irrelevant and unrewarding techniques), why not abolish it altogether? This was greeted with scepticism and sometimes dismay. Everyone was so *used* to an exam...

5

61

HEW

Yet nearly everywhere the English examination in the college of education had no function whatever, except to provide false disciplines and to tyrannise over the work. The only practical effects of it were to fail some students—at most 1 per cent—to award some distinctions, and to check on the rest. Where a few students were failed, it was mostly because they failed at teaching practice. Some failed in the exams, but their tutors usually knew beforehand that they would (unless they were cases of 'exam nerves'), and the simplest test would have revealed how unsuitable they were. In any case, early diagnosis of unsuitability is surely important in teacher training, because of the urgent need for places, and to avoid waste. It is surely not necessary to wait until a final exam or even one end-of-year exam to determine failure? The only function of the exam in the first and second years seemed to be to satisfy the examiners that students were fit to proceed to the next year, by testing the work they had done in the one just finished. But what tutor could not make such a decision about each student, from the files of terminal essays, and his other work?

Some tutors would argue that an exam counts as a whip—to help ensure that students work, do not waste their time, and cover the syllabus properly. But if we regard the essential discipline of the syllabus as that of training students to read and write well, then this is something that is inevitably known to the English lecturers from the weekly course of their work. If students are not gaining in the real work, then something else may need to be done. But all the exam does, as we shall see, is to obstruct this essential work, by encouraging bad substitutes for it.

Distinctions can surely be awarded on work done during the course? Here there is no need for an exam to help grade or select students for some future course of study. Of course, the problem may look different if a degree is to be awarded. But even so, it has long been argued in English that a 'piece of work', or a number of pieces, is a far more adequate test:* and in any case the disciplines of 'education' in English, as I shall argue, being concerned with creativity and whole experience, especially with children, ought only to be recognised for the award of a degree on assessment of a student teacher's whole educational experience and capacity, not on hasty paper-work in the exam room.

The more 'positive' its influence, the more importance it claims, the more limiting is the effect of the English examination. Inevitably an

* See F. R. Leavis, *Education and the University* (Chatto and Windus, 1943).

examination exerts control on the syllabus, and so students work 'to pass the examination'. In one college they actually practised 'timed essays': students were trained in beating the clock! Some lecturers argued that the students themselves like the 'challenge' of an exam. This kind of test is a challenge, they said—in the finished papers are samples of the best capacities fostered by the syllabus. Here the men and women show off their paces. In the excitement of the examination room, awed by the solemn prestige of printed question papers, foolscap papers tied up with pink tapes, with timed conditions of work, and the rest, they are keyed up, with sweaty palms, to give of their best. They will display the 'learning' they are to take away to offer children in school. But is this true?

The answer is to look at the product of the Eng. Lit. exam. Why is this so seldom done? Perhaps because it all happens at the end of the academic year, and doubts are forgotten during the holidays? Is it that we fear to disturb the illusory security which exams seem to provide? Is it because exams are carefully protected—as false comforters—by secrecy and 'confidential' memorandums? I propose to examine in detail the contents of some training college examinations.

Before we examine what actually goes on in an examination, perhaps we should enumerate briefly what the aims of a training in English are in detailed 'schematic' and practical terms, so we can then go on to ask if an examination really does test them. To make a rough list, our aims are:

(1) The development of a free, rich fluency, and the capacity to read well.

(2) The capacity to explore experience in an organised way by using words—to see connections in all aspects of living, and seek order in the inward life by verbal symbolism.

(3) The capacity to select material from experience and from books, including the deductions and opinions of others, and to organise them into expression of one's own. (Précis, 'research', etc.)

(4) The development of the sensibility, by drawing on creative sources of insight and satisfaction.

(5) The capacity to discuss creative works explicitly, with clarity, in one's own terms; to discriminate, and to relate the content of imaginative work to the experience of living, by criticism.

(6) The gaining of a sense of perspective in English literature and its place in civilisation, and a related acquaintance with critical writing.

(7) The capacity to teach literature—to be able to bring the excellent things

in literature (and related forms of culture) to the notice of others, and to foster their possession of creative works.

(8) The capacity to apply one's sharpest powers of reading and discrimination to children's work (which will, of course, require some knowledge of children).

(9) Some acquaintance with the social and political 'background' of English literature, of ideas and trends.

(10) Some acquaintance with popular culture, including folksong, and the development of a discriminating attitude to modern popular culture.

As we shall see, an Eng. Lit. exam is far from testing this whole range of capacities in English. What it tests is the ancillaries, the minor functions, which, because of their consequent exaggeration, can swamp the whole course. My point 9, for instance, grows out of all proportion, to the extent that you find students who cannot read and have no fluency who will pass an Eng. Lit. exam by noting fragments of a hazy perspective of 'Eng. Lit.'—without having read one work by some of the authors to whom they refer, as 'representative of the Age of Reason' or some such borrowed concept of general 'background'.

The disciplines enumerated above only have value if the capacities to read and respond, and to have a mind and voice of one's own, as suggested in 1, 2, 3 and 4, are achieved. In most of the examinations I studied, the candidates' work consisted almost entirely of generalisation unsupported by close attention to texts, to words, to lines of poetry or paragraphs of prose. And when students came to discuss actual words-on-the page they revealed a lack of the essential tools. They had not been trained in point 5. They did not possess critical terms of their own. They failed to possess these because they had not 'built' up terms for themselves, by seminar exchange, in which they had been asked 'What do you mean by "sentimentality"?' and so on. Most were not even able to use terms such as 'rhythm', 'texture', or even 'sound' or 'image'. So discussing children's own work relevantly—point 8—was utterly beyond them.

Yet this is the crux of the whole problem of developing new disciplines of creative English. If student teachers do not learn to read and discuss literature adequately, they cannot read and discuss, respond to and mark, children's writing. They will not be able to select the best for 'publication' by reading aloud and printing in class magazines. They will not know how to follow up children's imaginative interests as revealed in their writing. They will not be able to answer children's questions on poems: they will not know how to select from literature for children, at the timely moment. Yet this is what they will need to do daily in the classroom. The answers by students to a question on the child's poem below (p. 132) are most revealing—revealing as they do an essential incapacity to approach *creativity*. Will not this uncreativity in the exam be reflected in their teaching? For this I blame the influence of Eng. Lit. examinations on their work, not only in the college of education of course—but all the way up through the grammar school, too.

With such deficiencies it is hardly surprising that the students' sense of the relevance of their study of literature to popular culture and aspects of living in modern society was feeble (points 6 and 10). A training in English that is exam-bound simply does not cover efficiently all that needs to be covered, and tends to concentrate effort and attention on less relevant sections of the necessary work.

5

A TRAINING IN INSINCERITY

If we examine students' essays written during term, we find that the traditional 'Eng. Lit.' examination tyrannises over the whole syllabus.

The first essay in the folder I turn to is headed 'Samuel Taylor Coleridge—Lecture', revealing, in truth, that the essay is written up from the lecturer's notes. What a wasteful way to proceed! Surely the student could read the sources herself? This essay contains hardly anything but small detail of no great relevance to a study of Coleridge as poet. It might be relevant to the scholar, but even then it would be unlikely to show much illumination unless it had been done by an inspired critical-biographical writer such as John Livingstone Lowes, whose *The Road to Xanadu* shows a mind which can link 'scholarly' minutiae with the poem's text. Student 1* writes:

A perusal of the sonnets of William Lisle Bowles completely cured him for the time, however, of his metaphysical malady, but it was after the age of seventeen.

One's chief objection to such work is that such a girl *would not herself write like this.* She probably could not explain the word 'metaphysical' and only faintly grasps the irony in the word 'malady': the phrase is almost certainly borrowed by the lecturer from someone else's book, or taken from *his* lecture notes at college in his day—and so *ad infinitum.* The girl is being forced to write rather in the style of an old man—of a rather tedious scholar at long distance—biographical details which are irrelevant to her capacity to read *The Ancient Mariner* or *Frost at Midnight.*

When faced with the need for first-hand comment, such a student cannot rely on her rote-learnt scholarly language, and writes awkwardly in the extreme: here is the same girl's own voice:

'The Voice' was written after Hardy's wife's death, when he was a very old man. It is a superb poem, and poetically speaking, the death of his wife was the best thing that happened to him. The poem is written about his wife.

* To ensure anonymity I simply number students in the order in which they appear in this book.

The voice in the poem is the present, past and intermediate and in this way recalls the whole of the past. The introduction of his wife's ghost is cleverly done. The poem is nostalgic but returns to the present with a sudden immediacy in the last verse... *(Student 1)*

The banal awkwardness comes from the way a mock-scholarly language has been artificially imposed on a sensibility which is still somewhat raw. Yet one could, as she shows here, develop her own simple sincerity of response, if this were respected as that which matters most. There is no substitute for this.

When, later, we read this same student writing about Coleridge's private life, the effect is of a completely artificial performance in period costume and borrowed terms:

Throughout the spring and summer of 1795 Coleridge continued his lectures at Bristol but with his marriage on the 4th of October he withdrew from the eager intellectual life of a political lecturer to the contemplative quiet appropriate to the honeymoon of a poet... *(Student 1)*

This is ludicrously stilted and essentially insincere. The encouragement of such insincerity is harmful to the student's powers of literacy and to her training as teacher. No one should be encouraged to write such a phrase as 'eager intellectual life' until they know something of what the 'intellectual life' is—otherwise they are being encouraged to pretend—and we have enough of intellectual pretension. Even if they may refer to the 'intellectual life' they should hesitate before using a phrase such as 'eager intellectual life' as if claiming acquaintance with a range of intellectual milieux, 'eager' to 'not eager'. The epithet 'eager' is obviously borrowed—taken from notes taken from the lecturer who got it from Saintsbury or Herford, who perhaps borrowed it from a contemporary of Coleridge, and so on. It is inexcusable to encourage students—prospective English teachers—to reproduce the clichés of scholars at third or fourth hand.

Again, no young woman of this age would write 'contemplative quiet appropriate to the honeymoon of a poet': it is mincingly polite, coy, insincere, and not *hers*.

When she leaves the almost totally irrelevant biographical material for 'appreciation' of works, we find the terms of comment are borrowed, too—not only borrowed, but, as the challenge of 'unseen' literary analysis in other papers reveals, not even understood.

In poetic quality a keen sense of and delight in beauty, the infection of which lays hold upon the reader, are quite out of proportion to all his other compositions. The form is that of the ballad, with some of its terminology, and some of its quaint conceits both in 'The Ancient Mariner' and 'Christabel'. 'The Ancient Mariner', as also in its measure 'Christabel', is a 'romantic' poem, impressing us by bold invention and appealing to the taste for the super-natural...it is the delicacy, the dreamy grace in his presentation of the marvellous, which makes Coleridge's work so remarkable...In this finer, more delicately marvellous supernaturalism, the first of his more delicate psychology, which Coleridge infuses into romantic narrative, itself also then a new, or revived thing in English literature... (*Student 1*)

In such writing we find the language of a bogus discipline. The student has copied down a scatter of elevated phrases from her lecturers, who have copied them in their turn from some second-rate critic. She attaches them in her memory to 'Coleridge'. As we discover when she unloads them in an examination, she neither means them nor understands them. In her 'stock' answers all will be confusion: in her original 'unseen' analysis the terms are simply not at her disposal, possessed by her.

She herself would never talk about 'keen sense', 'delight in beauty', 'the infection of which', 'dreamy grace', 'lays hold upon', 'by his other compositions'. She may not know what 'terminology' means, and almost certainly could not explain 'quaint conceits'. She would never naturally use the phrase 'in its measure' (and perhaps should not be encouraged to). 'Bold invention', 'dreamy grace' are borrowed phrases which pretend a sophisticated critical apparatus where none is. She should be taught not to use words like 'marvellous', 'remarkable' and 'delicate' unless she means them—to encourage her to do so is to train that insincerity with language which is a characteristic of facile journalism. Such bad habits are a grave disadvantage to a teacher, whose concern must always be with sincerity of expression. 'Infuses', again, is not her word: and could she herself use the word 'psychology' with any confidence—would she have enough confidence to distinguish 'delicate' psychology from any other sort?

Compare her writing in this imitative way with the childish (and commendable) simplicity of a story she writes in her folder, *A Child in Summer*, or the pleasing directness of her first-hand analysis of some lines by Crabbe:

The little girl was wandering slowly along the dusty path parched by the heat of the sun and lack of rain to a fine powdery strip. The bare feet in their light sandals were streaked with the brown powder which clung lightly to the soft skin; the knees were black and the dress, which was white, was looking grubby. In her grimy small hands the child carried an untidy bunch of daisies and buttercups, some of which randomly slipped to join the dust at her feet. She never left them but immediately and slowly gathered them up again in an awkward movement. The arms were bare and brown in the sunshine and the hair, long and black and untidy; the face, the face of a child deep in some fantasy of her own little world; a brown face streaked with dirt...

(Student 1)

This is the directness of her own sensibility, and it is admirable. So, despite its rawness and hesitancy, is this:

The poet seems to be saying that although the fields are beautiful and colourful to the eye, they are really overgrown and untidy, and the beauty is only superficial. The very last line of the poem conveys something of this:

> *And a sad splendour vainly shines around.*

...The rhythm and rhyme scheme of the poem is rather monotonous, but is an asset the the poem as it helps suggest the sameness of the fields. All the land is overgrown and looks the same...

(Student 1)

Immature as this writing is, it is the real thing, and infinitely to be preferred to the imitative pomposity of her writing about Coleridge. The implication is obvious: if students read poems, discuss them in their own voices in a seminar, and then write about them—then we shall have a valuable 'voice' growing in them. To copy 'notes' from a lecturer and write these notes into essays is by contrast a false process, because it bypasses true engagement in favour of 'acquiring information about'.

Of course a student needs to read good 'literary' critical writing. By this, naturally, I do not mean the kind of biographical stock account of a writer found in the literary histories. But even if a student reads a good critical essay, there seems to me no justification for his reproducing such material as his own 'essay'. Again, this is not to deny that students will inevitably, and especially at first, tend to reproduce something of the work of others (adult writers do it often enough). There are two kinds of indebtedness. A student may well be impressed by and influenced by an important critic, and this will show. But he can show himself to be actively assimilating the influence, making it his own thought, and thinking it through for himself. This is not the same thing as serving up cold something which has not affected his own mind. There is no justification for simply 'writing up the lecturer's notes', nor for the dishonesty of writing about works students have never read. The assumption underlying this procedure is bad—that the students' own responses are not valued, while they are encouraged to get away with impostures.

If they simply copy and repeat, the students are not being taught sufficiently to think about the critical terms, the words they use to discuss works of literature. Yet when they are in the classroom, this is what they will need to be doing all the time. There, I am sure, many failures must be due to the fact that the teacher has been too much cowed by printed authorities and taught to reproduce as his own superficial comments from poor literary histories.

Whatever will the next student say in the classroom about the Romantic poets and their attitudes to Nature? In her essays she is sentimental and vague: she misses the best in Wordsworth, applauding some very poor examples—because she has never been left to browse through Wordsworth, under pressure from challenging seminar discussions of his poems. She has perhaps 'done' Wordsworth in a week without really reading him. So she writes about passages in words which patently are not hers:

When the simple vision is seen simultaneously with the majestic, the experience is intensely powerful and almost religious... *(Student 2)*

One wonders what the source of such a sentence is. Compare it with another piece of an essay in which she speaks in her own voice, again about Crabbe:

In the thistles which choke the roots, the poet sees an image of something terrible to come: the spikes of the thistles represent the spikes of guns and bayonets: *And to the ragged infant threaten war...*

Poppies, also, seem to represent the fatigue and apathy of human life, as nodding, they *mock the hope of toil...* *(Student 2)*

This is simple, and there is no pretence about it: it is good. It is hers, refreshingly first-hand. The simplicity might alarm her examiner, because it is so unpretentious (to accept that one's work is at such a naïve level might threaten a loss of status?). But it is a genuine response to art—and so something for a tutor to be glad about, for now a true training in critical analysis can begin. For this is how she will speak when she is teaching poetry in the classroom—and the pupils will be able to understand her. The borrowed mouthings about 'the majestic' only encourage a pomposity that makes her writing elsewhere awkward and insincere:

As the novel is almost entirely a portrayal of Emma's character and the consequences of her impulsive actions, it is fitting that Jane Austen sets the first chapter at Hartfield, Emma's house, where Emma's true character can be revealingly contrasted by the gentle humour of her father and the dignity yet gentle tolerance of Mr Knightley... *(Student 2)*

An examiner noting the tone of this ('it is fitting that') must surely think to himself that this is a second-hand essay memorised almost by heart (see the revelation of these processes below, pp. 95 ff.). A student can write like this, and often does, when the book in question has roused no interest in him whatever: so what value has it?

In her essay on Wordsworth, this girl student reproduces attitudes to Wordsworth's 'moral' development which are derived from critics. They cannot have been the best critics, for those would surely have thrown her back on her own responses? It is perhaps because of this that the better critics such as F. R. Leavis, the William Empson of *Seven Types*, or L. C. Knights, tend to be submerged by second-rate writers whose generalisations are applicable in any vaguely relevant circumstances (e.g. 'Wordsworth sought consolation in nature'). Such second-rate criticism tends to suppress the disturbing truth of art. So, it is eminently 'quotable' in exams, while those critics who make the closer and more revealing analyses are difficult to abbreviate (though T. S. Eliot, who does not stick so close to texts, can be—and is an obvious exception—though his best work is travestied by conventional restatement). In her work student 2, while failing to find a Wordsworth who touches her own feelings, learns to write in an insincere mode of 'appreciation' about something she does not in fact appreciate. No wonder one picks up from such work a flavour of desperate boredom. This kind of effort bears no relevance to work in the classroom:

These experiences are not explained merely as the reaction of a guilty mind, but as an example of the moral chastening and brooding dominance of nature... *(Student 2)*

On *The Ancient Mariner* student 2 begins to give us something of herself—critical phrases she understands. But even so it is disappointing that of such lines as

> Her beams bemocked the sultry main
> Like April hoar-frost spread...

she can only comment:

There is something horrific and unreal about what is usually a beautiful sight: the moon on the water. Coleridge's moon holds a supernatural beauty. Under the influence of the supernatural effects the Mariner's attitude changes...   (*Student 2*)

The girl could have given something more originally hers, have been forced to read, to assess and express, had she been obliged to discuss the exact effects of the words 'bemocked', 'sultry', 'April', 'hoar-frost', in their context. But to have done so, from choice, would require a different kind of training from that she has had on Wordsworth and more like that she has had on Crabbe. This, besides being better teaching and a better critical approach, would equip her for the classroom in a way such gestures as hers about 'supernatural beauty' would not.

Exactness in discussing meaning is the mark of good criticism. I found it shocking, in so many examinations, that students' work showed so few signs of a training in adequate critical discussion. I found it alarming that one could predict that a student who quoted a line from Keats such as

> The murmurous haunt of flies on summer eves,

or the last stanza of the great *Ode to Autumn*, would follow it parrot-fashion by some cliché remark borrowed from a textbook of literary history, such as: 'familiar yet beautiful images of Nature' or 'meditation of sensuous beauty'—and never say anything fresh of her own, about flies or the moods of summer, or even onomatapoeia.

There is surely little point in discussing Keats if one does not discuss the *words* of his poetry. Keats's attitude to Nature is indistinguishable from the way his art re-creates in sounds the tangible presence of natural vitality, as by the consonantal muscularity which enacts richness in the mouth in such a line as

> The grass, the thicket and the fruit-tree wild...

Such a line we can use to discuss enactment, and the tangible exactness of imagery and movement. The student I am quoting does not even point out the noise of the flies in 'the murmurous haunt of flies', or the enactment of the gathering swallows, their sound and flittering in 'gathering swallows twitter in the skies'. Such sound and rhythm, in its textural exactness and beauty, by involving the sensations and muscles of the mouth (even if read silently) is capable of promoting a relish for vitality. The words themselves convey a glad grasp on experience: and thus impart an enrichment to the inward life. It is possible to convey this to children, but only by discussing words and meanings, not by reproducing the comments of literary historians.

To follow such close criticism as that of F. R. Leavis on Keats's *Odes* would have promoted exact attention to meaning. But Leavis's most important essay, full of 'felt life' itself, in *Revaluation*, was not in the book-list of the girl whose work I am discussing.

The essays I have been discussing are pieces of work done in term. No wonder that when it comes to the examination we have on the one hand discussions of verse in detail which flounder and are totally inadequate, and on the other generalisations about Romanticism which are often so vague as to be meaningless, and at worst howlers.

6

PROCESSED UNLOADING

Sometimes as examiner I make tables showing the proportion of students answering each question in each examination paper. The examiners tended to set essays they had required during the term: I insisted on some alternative 'unseen' exercises in close critical analysis. The result was what one would expect—the students funked these, and picked on those questions to which they could reproduce a stock answer from memory. They had practised timed essays, and unloaded these in the exam: it would have been just as useful if they had learned off passages from the *Koran* or the *Karma Sutra*. So much for 'covering the syllabus': it often means no more than stocking up a dozen essays on 'safe' subjects, in the hope that at least six will turn up in the question paper.

There might be some point in such practices if the students had been forced to take an external examination for which the staff had to do their best to prepare them—dictated, say, by some distant university. But in one of the exams I am discussing the staff of the college of education imposed the ritual themselves. They practised the students in timed scribblings, then retreated to the role of examiner, and tested them in this futile activity. The startling thing is that lecturers devise and maintain this system themselves—and do not see what is happening. Because examinations are so much part of our educational experience, and because they help to obscure the disturbing problem of aims, no one questioned the value of this process, even though the evidence that it was a mockery should have become clear to them in the marking.

The aptitude chiefly tested by the examination is that of studying sources, taking notes and composing them into the terminal essay. These essays are memorised in 'revision', and then scribbled out as well as may be against time in the examination room: if not always without texts, mostly without time to look at them. In this mug-and-scrawl performance the students save energy and time by reproducing in the examination room the 'safe' stock essay, and gain their marks on these. In one college, in examination of the first year the stock answers were chiefly these:

74

The Romantic Movement   Matthew Arnold's poems
Wordsworth's *Tintern Abbey* Swift's satire
Keats's *Ode to Autumn*   *Richard III*, Act I, Scene 2 or Act III, Scene 4
Clare's *Thrush's Nest*   What happened to English poetry from Pope
Hopkins's *Collected Poems*★   to Wordsworth?

It became apparent that it would, here as elsewhere, be possible for a student taking either a general English course or the literature papers to do well in such an examination without doing anything more than reproducing by memory from his notes what his lecturers had more or less said during the year on nine such subjects. In many places the syllabus was hopelessly large, but yet students could do well in a few safe stock items. With many authors they could pass without reading the books—so long as they remembered the general drift. Of course many students tackled other subjects, and wrote with knowledge of the texts, some of them bravely undertaking 'unseen' passages for analysis.

Here and there a good tutor will get students to write poems or short stories in the general English examination, and at one college where George Crabbe had been well taught many students tackled unseen lines from his poetry for analysis.

But what happened to these braver students? As we shall see, those who took the plunge into first-hand analysis put themselves at a disadvantage with the examiners—who tended to mark them down. Naturally their efforts tended to be callow, groping, and lacking in facility. They stumbled, were unsure, and tentative. But in effect they were *penalised* for reading with open responsiveness, reticence, and doubt—the very qualities we need to stimulate, but against which an examination, because of its very nature, manifests. The most telling objection to examinations is not merely that they encourage the cheat, and the facile stock answer, but that examiners tend to put the sensitive and responsive spirit at a disadvantage because such a student will not mouth hypocrisies, must ponder, and tends to write with a direct openness to literary experience that looks—at a brief glance in marking—like ignorance. The best examinee is often the man least troubled by conscience and creative doubt.

As I have said, we are not concerned with the problem of social fairness to candidates—making it possible for them to 'get the best

---

★ Hopkins was added to the examination papers by me at the last moment, and the question was obviously chosen by several students because late work had been done on it. The subject was still in their minds, and they twigged that the panic about Hopkins meant they were going to get Hopkins.

opportunities' and so forth—in examinations in colleges of education, since they do not count in the outside world. The examination only counts, as I have suggested, in a minor way. Yet despite its lack of any real justification, colleges are prepared to accept the effect of the examination on the whole previous year's work in English. I shall try to show that the central discipline of this kind of examination is the capacity to regurgitate a number of stock essays by heart and that this is inimical to all that a young teacher needs.

Here I shall take three examples of the kind of stock answer required by this exam work. Firstly, I shall reproduce a number of paragraphs from a stereotyped answer on 'the satire of Swift' attempted by nearly every candidate in an 'English for All' paper. Secondly I examine another question of a similar kind on a stock subject (English Poetry, Pope to Wordsworth) where, obviously, 'something went wrong'— not enough time was given for drafting and memorising the stereotype, or the students have merely reproduced half-remembered stereotypes from their grammar school days. Here the consequences can be seen to be quite disastrous, a complete travesty of English studies, at times painfully ludicrous. Thirdly, I shall examine stereotype drafts written in term-time by good students and compare these with what they reproduced in the exam room. One stereotype itself is inadequate and vague, even though it was awarded the mark 'B'. My purpose is not only to show how much it loses in the exam room, but that the memorising and reproducing do nothing for the man—they only reinforce the processes by which literary comment hardens into cliché. He merely learns his own indifferent essay by heart—a sheer waste of effort and time. Since the examiner knew the original essay, what do either tutor or student gain from spending time on a poor reproduction of a poor essay in an exam? Things are no better when a good essay is reproduced from memory. I hope these analyses of the content of such examinations expose once and for all the inefficiency and inimical influence of the Eng. Lit. examination, certainly as far as training teachers of English is concerned.

First, then, the question on Swift's satire. The question on the paper was this:

Describe, as precisely as you can, from your reading of *Gulliver's Travels*, Swift's technique in his satire.

When I first saw the proofs of the examination papers I could see that the lecturer had put this in as a 'safe' question. The students had all done their essay on it. Behind the 'safety' was the conventional assumption that Swift is an appropriate writer for school use, providing the supreme example of 'satire', this being one of the 'forms' of literature to which English students should be introduced. The roots of this approach, of course, go back to the early standardised forms of Eng. Lit. teaching: we should study 'form'—comedy, satire, tragedy, the elegiac, the epistolary style, and the rest—as if studying literature were like studying bakery. Form and style: all we need to do to be literary people is to learn a series of modes of employing the flour, water and baking powder of language in various combinations; and Swift is an obvious choice as a plain example of 'satire'.

Faced with such a question on an examination paper as external examiner I feel baffled. In an obvious way, yes—Swift's writing is satire. But such an approach to Swift seems to me crass: the concept of 'satire' and form, I can see, will get between the student and the writing, tending to blanket a complex—and fascinating—case. 'Satire' suggests genial fun, and a politeness of tongue-in-cheek—such as is utterly inappropriate to Swift's savagely destructive energy. Indeed, the point is that we need to question whether Swift's effect *is* that implied by attributing to him a 'satirical' intention. Such a label implies that this writer knows what he is doing, and that we accept that he knows what he is doing—he is consciously using 'satirical form', to expose social injustice, political corruption, and so on.

The truth is that Swift did *not* know what he was doing, and obviously knew less and less as things went on—his life-denying energy takes over, until there is no question of the adoption of a deliberate stylistic device to make points that could not be made otherwise. The schizoid phantasy takes over, and there is an intense dynamic which seeks to attack and seize the reader—and involve him in a destructive dissociation. Or this is my feeling about Swift: certainly I would want students to test their responses to him, to see if they agreed, without being fobbed off by the polite label 'satire'.

There are two ways of approaching Swift. One is by the naïve, unconditioned response of the first reading—in which a student may be asked, 'How do you find this? Does it square with your own experience —your own picture of man?' The other is to consider which accounts of Swift in literary criticism seem to illuminate his work. How does it compare with *Animal Farm*, or Rabelais, or *Erewhon*, or Sir Thomas More? Such an account, of course, cannot leave out F. R. Leavis's essay on *The Irony of Swift* in *The Common Pursuit*. Ever since this appeared we need to approach Swift's work with care—and especially to decide with some deliberation whether Swift *is*, as is commonly supposed, good for children—or, indeed, the creative satirist he is customarily taken to be.

The points Leavis made are crucial, and indicate a need to explore beyond the bounds of literature at times (in a way that surely links with psychology?).* I feel sure that, so long as they are not misled by conventional judgements, students will discover for themselves the destructive *frisson* in Swift—and could link it with the same exploitation of fear and disgust in some modern writing.

Hypertrophy of the sense of uncleanness is not uncommon; nor is its association with what accompanies it in Swift. What is uncommon is Swift's genius and the paradoxical vitality with which this self-defeat of life—life turned against itself—is manifested. He is distinguished by the intensity of his feelings, not by insight into them, and he certainly does not impress us as a mind in possession of its experience...†

Our work in literature is surely directed towards 'insight', and towards 'the possession of...experience'? Our quest is to enable new generations to come to terms with the more troubling experiences which life can offer, through art. The paradoxical feature of Swift's writing is that his self-assertive energy manifests against the acceptances of the fleshly and mortal 'Yahoo', in favour of the void 'reason' of the bodiless Houyhnhnm, in a schizoid way. With the Struldbrugs and Brobdingnagians there is an obsessive preoccupation with physical grossness and a fear of bodily life, that brings us to the verge of psychopathology, and the utter substitution of hate for love.

Swift's case, then, brings us to the edge of neurotic disability—not less a problem to many in the modern world than to him—the inability to accept certain features of mortal existence. But what did he make of the problem? We need to approach Swift with diffidence, and perhaps the best tack, since Swift is undoubtedly 'there' in the canon, is to ask the students themselves, in the first place, what they think of him. This begins with the effort of pondering the meanings of his words.

---

* There is a most valuable essay on the unconscious origins of Swift's Gulliver phantasies by Sandor Ferenczi. Ferenczi discusses the evident fears of annihilation Swift focuses on a preoccupation with the genitals and terror of castration (see Sandor Ferenczi, *Final Contributions to the Problems and Methods of Psychoanalysis*, Hogarth, 1955).

† *The Common Pursuit*, p. 85.

For such reasons I added an alternative question to the papers, choosing a passage in which Swift exploits the evocation of disgust and recoil, with active energy, and with that vitality which asserts the negation that Leavis notes:

Either
(*a*) Describe as precisely as you can, from your reading of *Gulliver's Travels*, Swift's technique in his satire.
Or
(*b*) Study this passage:

'One Day the Governess ordered our Coachman to stop at several Shops; where the Beggars watching their Opportunity, crowded to the Sides of the Coach, and gave me the most horrible Spectacle that even an *European* Eye beheld. There was a Woman with a Cancer in her Breast, swelled to a monstrous Size, full of Holes, in two or three of which I could have easily crept, and covered my whole Body. There was a Fellow with a Wen in his Neck, larger than five Woolpacks; and another with a couple of wooden Legs, each about twenty Foot high. But, the most hateful Sight of all was the Lice crawling on their Cloaths: I could see distinctly the Limbs of these Vermin with my naked Eye, much better than those of an *European* Louse through a Microscope; and their Snouts with which they rooted like Swine. They were the first I had ever beheld; and I should have been curious enough to dissect one of them, if I had proper Instruments (which I unluckily left behind me in the Ship) although indeed the Sight was so nauseous, that it perfectly turned my Stomach.'

From such passages some critics argue that Swift was unhealthily obsessed with the uncleanness of bodily life, and is too sick and negative a writer to be given to children. Give your opinions, with textual analysis, of this and other passages.

Only three students out of sixty-six answered this extra question, and only one answered it adequately. While he saw that Swift has an obsession with uncleanness, the conventional concept of Swift's 'satire' given him by his teaching does not allow him to do anything but make excuses, to defend Swift against criticisms of his negative elements:*

I do not think that Swift was, as some critics argue, unhealthily obsessed with the uncleanness of bodily life. I think it is true to say, though, that in his desire for detail, with which he embroidered this 'travel' story, he was carried away into unnecessary length...It is as though Swift had studied the human

---

* A similar process was at work in a student who, discussing *Jane Eyre*, managed to expose its unrealities—its puppet figures—but went on to seek to excuse this criticism as 'not mattering' with a 'work of genius'.

body and habits through a powerful microscope and seen these things, which, enlarged, horrified him. I think he overemphazid such passages into horrifying his readers into believing the story... *(Student 3)*

This student, using his own words, is on the verge of perceiving how Swift needs to involve the reader in the credibility of his neurotic distortions: he goes on about the maids of honour in the Court of Brobdingnag:

He describes the maids, not as a tempting sight, but 'giving me any other motions than those of horror and disgust'. Gulliver is sickened by the coarseness, unevenness, and variable colour of their skins... *(Student 3)*

But the student excuses Swift by explaining that 'such matters [as Gulliver urinating on the palace fire] may not have been thought obscene and the people of the day *were less inhibited* than we are' (my italics). This essay would at least have provided a useful starting-point in the discussion of Swift—and led to consideration (which none of the three students proved able to do in tackling the question) of the words and their activity. The generalisations about satire would have a good shaking up, over such phrases as 'their Snouts with which they rooted like Swine', 'dissect', 'nauseous', 'turned my Stomach'—the negative activity being so vigorous that it cannot be checked, and progresses obsessively and aggressively forward in its offensive sensationalism. But the effect of the polite critical notions of 'satire' effectively screens (or perhaps the word is 'castrates'?) any vigorous approach to the meaning of language. Meanwhile, it might seem, Swift is to be enlisted in the category of 'uninhibited' writers such as appeal to fashion today.

Even this inadequate answer, in his own terms, by this one student is much to be preferred to the stock answers given by the other sixty-three students who relied on this question as one of their 'safe' options:

Swift is known to have told Pope that his aim in writing 'Gulliver's Travels' was 'to vex the world rather than divert it'. For the most part, however, his satire is gentle and is usually aimed at the political scene in Europe. It is only when we reach Book IV that we discover that here Swift is satirising the human race as a whole and not individuals. *(Student 4)*

Gulliver's Travels is clearly a satire; that is, it exposes the weaknesses of mankind by holding them up to ridicule. But who and what were its objectives? Swift states his own view very unequivocally. 'I thoroughly hate and detest that animal called man although I heartily love John, Peter and

James'. He set out to prove the error and falsity of a philosophy of optimism. He also stated that 'Upon this great misanthropy...the whole building of my travels was erected' and he adds that they were written 'to vex the world not to divert it'...

(*Student 5*)

This student's answer is perhaps the most competent of the stereotypes. At least she understands much of what she is writing, even though it is pathetically unoriginal.

Swift wrote 'Gulliver's Travels' 'to vex the world rather than divert it' and he achieves this end by his employment of satire. He admits himself that he 'had too much satire in his vein' and this is extremely obvious in all four books contained in 'Gulliver's Travels'...Swift did not despair of man as an individual, but rather felt that society, mankind as a whole, was degenerate. He especially considered politics and the court of George I and Queen Anne as corrupt, and also had angry thoughts on the dissention between the high and Low Church in England. All these are subject to his satiric vein...

(*Student 6*)

No student protested against 'Swift's assumption that man is degenerate'. The trouble with this kind of academic teaching is that it generates a 'new aestheticism'—a split between art and life. The student fails to engage in the creative personal process of accepting or rejecting the phantasy experience, in testing it against his own experience. To such students Swift does not 'mean' anything that bears on their own experience or attitudes to life—it is 'apart' as 'literature', and they do not even respond to Swift's words:

In 'Gulliver's Travels' Swift sets his satire in the form of an imaginary voyage, giving it a solid core of intellectual meaning. His chief aim was to vex the world rather than divert it. In a letter to Pope Swift wrote that when the world was ready to hear it or 'rather than a printer shall be found brave enough to venture his ears, the chief aim I propose to is vex the world rather than divert it'.

Swift's satire was hard and shrewd. Logic was a formidable weapon...

(*Student 7*)

Again, such 'knowing' phrases as the last do not proceed from the student herself, and it is very unlikely she could begin to substantiate them. They are half-remembered fragments from lecture notes and books, and nothing more—ghosts of the orotundities of dull essayists.

The technique adopted by Swift when writing Guilliver's Travels was that of satirical ironic prose. In this book Swift satirises life the late seventeenth and early eighteenth century. He himself said that his aim in writing the book was 'to vex the world rather than divert it' and it was by means of satire that he achieved his aim. (*Student 8*)

A student who departs from the stereotype and approaches the subject in her own language and terms is given in the marking only one point above the others:

The techniques which Swift employed in his satire were numerous. The main ones, however, which can be seen in each book are his use of contrast—that is of constructive satire being contrasted with destructive writing; his use of scale; and his use of Gulliver's innosense.

In the first book, for instance, we have destructive satire of eighteenth century England which comes in the form of the corrupt practices of the Lilliputians. The jostling on the political ladder of England, is here reduced to a tightrope walk, upon 'a slender white thread'. (*Student 9*)

Inevitably, the student who departs from the memorised phrases of the lecture-note stereotype finds herself floundering in a not very adequate critical idiom of her own, and writing rather clumsily. But student 9's effort above stands out even because of her stumblings, above the piles of clichés:

When Swift was writing Gulliver's Travels he said that his aim was 'to vex the world rather than divert it'. He was attacking the court of George, and the Whig administration of Walpole... (*Student 10*)

In a letter to Pope Swift said that the aim of his writings was to 'vex the world rather than to divert it', however it hard to believe that what did vex the world could hardly not infinitely divert it. In Gulliver's Travels Swift was trying to prove how silly were English (and indeed European) methods of running politics, economy, and society in general. Therefore he wrote a book about societies completely different to that of his own and attempted to prove that they were better than the English one. (*Student 11*)

Here I pause to discuss the effect of such literary oakum-picking on the student's capacities to deal with words and children—a question sharply raised by this particular candidate, because we can compare her writing about Swift with her writing about a child's poem. When students have been exposed to the false and irrelevant discipline of learning such critical platitudes parrot-fashion they are cripplingly inhibited by this training when they come to approach the work of children. Of what value to student 11, as a prospective teacher—for her profession that requires her to be a more than adequate reader—is her work in this essay on Swift's satire? Her clumsy gestures at the 'background' of Swift is the complement to her answer to a question which presented students with a child's poem and asked the candidates 'what they would say to the child' (for the poem see below, p. 131). Student 11's answer on Swift is the corollary to her inability to approach the problem of creativity and its encouragement in this child. This is the best proof I could offer of the anti-creative effect of the work thousands of student teachers are doing in the name of Eng. Lit.

I think that this poem is very good for a child, but that a girl of twelve ought to have some ideas of the significance of punctuation. The poem has extremely good substance and with help could soon be improved with only minor alterations. The primary thing to do in my opinion, is to correct the spelling first of all and insert a minimum of punctuation to give the poem more sense than it already possesses in its present state.   (*Student 11*)

We shall see later how this failure to be able to receive creativity from children is the price paid for the conventional 'Eng. Lit.' training. It turns out Philistines. Is it any wonder when a whole batch of answers begin with the stock phrases muttered by mnemonic patter on the way to the exam room ('Swift's intention in writing *Gulliver's Travels* was "to vex the world rather than to divert it"...', 'Jonathan Swift is known to have said "I hate and detest that animal called man"...')? Even the best of the answers showed the students engaged in an activity which is concerned with no more than dashing down minimal precipitates from the memory of lecture notes, perhaps a few sources, and hardly anything from first-hand reading of the creative work itself. As such their answers give the vaguest picture of the author—a *Reader's Digest* summary of a complex, difficult, disturbing and malevolent body of work. Such answers as the following could be given by a computer:

Swift's satire is sometimes mild and humourous, sometimes merely contained in a very concise phrase, but by the time of the journey to the Houyhynnm, it has become bitter, very pronounced and rather repusive. Although at the time of publication Swift's book was accepted as a book of fact, I believe that his intention in writing it was much deeper than this—Swift was rather using his literary talents to give his idea of eighteenth century England, as he saw it.   (*Student 12*)

This attempt secures 25 marks (out of 33), and is described in the margin by the examiner as 'Good work. A full answer. Competent technically, and with some sound ideas about the book.'

But the paragraph just quoted bears so little relationship to Swift, whose intention was certainly not to 'give his idea of eighteenth century England', that it is meaningless, and futile. Was *Gulliver's Travels* ever accepted as 'a book of fact'?

In such answers literature is sterilised and the guts taken out of it. Swift has been made conventionally 'safe' for a 'safe' examination answer: what we have here is Swift at third or fourth hand, seen through a glass darkly, hardly worth attention, and recommended for the 'picture of eighteenth century life'—whereas, to the reality he deals with, he is but a partial, uncertain and destructive guide—and the attention to 'outer' social matters is but a disguise of an attack on our inward life. Not even the best Swift is seen, with his strange energy of distorted vitality. There is no local analysis of words, of the tricks and surprises this strange writer springs on us. Indeed, there is no real interest in Swift, in most of the sixty-three stereotype essays. The students had their texts of Swift in the exam room with them (on my insistence) —but none of the candidates used them to quote from.

The most disturbing proof that the effect of exams is inimical comes when a question intended to offer an opportunity for the display of stereotypes 'goes wrong'. The spectacle reminds one of a Pavlovian experiment in which signals are switched and the animals become confused and 'neurotic'. I shall now examine answers to 'What happened to English Poetry from Pope to Wordsworth?'. At the special request of the examiners this question was retained in a question paper, despite my recommendation that it should be deleted. At least I provided an alternative to it—a close analysis of Wordsworth's lovely sonnet *Surpris'd by Joy*. Of the seventy-odd candidates only five tackled this poem. Perhaps it was too hard for them, though the poem usually makes a direct impression on members of an ordinary extra-mural adult class. There is nothing obscure about it, certainly.

But the general question provoked some incredible statements about literature:

Poetry in one form or another has existed ever since there was an interest taken in literature. In the Middle Ages there were the ballads—men travelling round the countryside informing the people in the villages and towns of the news, relating it in verse. Then we get the classical plays at Ancient Greece and Rome, written in verse using variations of the ballad form... The years 1700–1800 were years of turmoil and hardship for the people of Britain. Life was formal (seen by the dress-crinoline and the formal dances)... [Pope] shows how they attend to trivial things in everyday life...When we come to Crabbe we are presented with the poverty and discontent of the working classes...We must not forget the supernatural bent of Coleridge when it is obvious that poets were looking for something better in life and were reverting to the past and the simplicity and straightforwardness of the Middle Ages.

The final touch comes with Wordsworth and his presentation of man and everyday life.

(Student 4)

The poetry of Pope and his contemporaries is ordered and peaceful. Throughout the late seventeenth and early eighteenth century poetry spoke about beauty, the pleasant countryside, cool glades, and most import, sunshine... To Wordsworth, the peaceful glade and pleasant afternoon were to be shuned in favour of crags...Wordsworth dispenses with regular rhythm...

(Student 13)

Pope and Wordsworth lived in different eras when the questions of religion, politics and man's right did not have the seem meaning. When there was such a great difference in their environment it is obvious that this difference must show itself in their writings.

Wordsworth, having witnessed the French Revolution, the Industrial revolution, the fights for the franchise, education, slave trade and the burning liberal awakening of the nineteenth century, wrote in a revolutionary manner...He had a pantheistic attitude towards nature—for him it almost took the place of God. He used the natural beauty of nature often to create a mystical element. His work is often coloured with a mediaeval style and content—mystery and the supernatural...Pope's extract from 'Windsor Forest' is one of his best works. During Pope's life, eighteenth century England was still an agricultural country. The picture of Paradise is the typical Dresden shepherdess. Men were not regarded as equal in Pope's time. The upper class or landed gentry were the important and top people. Pope's work reflects this C18 attitude towards life. The appeal is not wholly to the heart, which has not yet been liberally stirred, but is also to the ear ('frozen thunder') noise of gun and to the eye ('purple crest') etc. The setting, early morning in Windsor Forest is typically upper class. Pope shows his pity for the under class by showing his pity for the birds.

they feel and leave their little lives in the air.

The change in life and poetry style was inevitable. Blake forcast that it could not be stopped in 'The Tyger'...Blake calls the industrial revolution 'The Tyger'...

(Student 14)

In how many establishments are young people like student 14 scribbling in panic their way through such nonsense, without interest or engagement in what they are doing, in the name of Eng. Lit.— as an examinable subject? Is this the new depth to be made possible by a three-year course? Is this work of 'university status'—scratching down half-remembered scraps from lectures?

Of course, much the same kind of thing goes on at the university: we all did it. It is significant that another question in the same paper, 'Say how you would set about interesting a young friend in poetry, discussing actual poems by which you would hope to tempt him (or her)', was attempted only by one rather poor candidate. The students simply had not been interested in poetry, nor, really, in literature in general. Only sometimes, under the discipline of work, does something spark and come alive (see below, pp. 116 ff.). The reason is not far to find: the students do not read anything, or share readings with their tutors. The 'background' or 'historical' approach to literature not only tends to treat it as a body of dead artefact, creative works as clues to the 'conditions of men', in past centuries—it becomes a substitute for reading the works. There simply is not time to read them. So we have the school-report biographies of Coleridge ('At the age of nineteen Coleridge left school to attend Jesus College, Cambridge. Although very studious—within a few months of his entrance he won the Bourne Gold medal for a Greek Ode on the Slave Trade—his reading was desultory and capricious'). We have the false voices of borrowed literary history ('the spectral object, so crude, so impossible, has become plausible, and it is understood to be but a condition of one's own mind': when did a young man of nineteen ever write thus?). There is a kind of activity going on—and the exam seems to be a guarantee that something has been achieved. But this external activity is a mock, and merely disguises a total absence of that engagement of interest which alone can bring literary studies alive—alive in anyone, let alone those who are going to become teachers, presumably destined to inspire others.

The reader may consider the effect on the English sensibility of writing as these students do, in answers to this question:

During the century between Pope and Wordsworth there was a change in English poetry. Pope, in the eighteenth century, was typical of the classical era of poetry. He was writing at a time of war when people wanted calm and security...in his poem on 'Windsor Forest' Pope is directing his words at the general public...an underlying moral...that we should not fight and kill our fellow men...Wordsworth was also writing at a time of war and revolution, but his writing is typical of the romantic era. His poems were to appeal to the emotions. He wanted people to think after they read his poems.

His poetry was centred around the ordinary man and woman...There is much more use of metaphors, similes and symbols than there was previously. Sometimes his poems verge on the sinister... *(Student 15)*

An afternoon spent reading Pope's *Moral Essay* on Sir Balaam would confute the statements made here about the poet. They could not have been made by a girl who had ever been brought to respond with pleasure to one passage by Pope or one by Wordsworth.

The stereotyped question on 'English Poetry from Pope to Wordsworth' reveals how disastrous such wild sweeps of 'background' perspective are, if they are based on accumulations of notes and little reading. One can see what the examiners were at. Attitudes to nature, man and society changed: so did the poetry—in its content and form. From the notes the phrase sticks, and is repeated, that 'Pope reflects the desire of the people to live a quiet, more ordered, civilised secure way of life'. The students flounder, to link the French Revolution with the changing attitudes to nature, from a taste for 'quiet glades' to one for 'crags'. The focus of students' attention is apparently Pope's *Windsor Forest*, on which no doubt the lecturer had a *pièce de résistance* in his repertoire. (This was often followed by reference to the change from iambic pentameters to blank verse, the course of events and philosophies in France, the industrial revolution, and then *The Prelude*, which apparently presents man as 'a lonely oarsman battling against the elements of a rough and rugged nature'.)

Wordsworth did not need classical references, his work was made real by his choice of ordinary words and by the accentuation of certain syllables as:

/ / / ✓ / ✓ ✓ / ✓ /
Rose up between me and the stars and still
✓ / ✓ / ✓ / ✓ / ✓ /
With measured motion like a living thing,

the accentuated syllables at the beginning of the line portray a feeling of shock and then the second line with its regular rhythm is as if the cliff were walking or getting up. *(Student 16)*

The student receives a tick in the margin for this attempt at rhythmical analysis, no doubt inaccurately reproduced from his notes. If / indicates a strong stress and √ a light one, it is impossible to read the lines as he marks them. As so often, when it comes to actual close reading the students reveal that they cannot do it.

This analysis of Wordsworth was followed in the stock answer by an analysis of Gray's 'elergy' and then by one of Goldsmith's *Deserted*

Village. There was hardly any evidence that the students had read the verse to which they refer. Much of this poetry requires effort to read; and for this activity half-learnt fragmentary summaries of notes are no substitute. No wonder the answers are so bizarre:

In the time of Pope classical literature was conventional. In fact, every part of culture was influenced by Classicism and convention, with no thought of the individual common man, but society as a whole was prevalent. Such poets as Gray are completely typical of this period as is seen in his 'Elegy in a Churchyard' which is written in the heroic couplets typical of the age, and showing in such lines as 'The curfew tolls the knell...'.

...The influence of classicism is one form of Dante's influence. The beetle mentioned in the poem is further evidence of classicism and unoriginality, having often been mentioned by such prominent poets as Shakespeare. There was severity in all aspects of literature, adherence to strict metre, strict subject, strict language...The Romantic Revival began as a protest to this conventionality and developed in complete contrast to the Alexander Popian poetry... *(Student 6)*

Compare this same student's stereotyped answer to Swift above (student 6, p. 81): whatever can a year's 'literary studies' have meant for her, poor girl? Yet she gave elsewhere a reasonable account of the wooing of Lady Anne in *Richard III,* and is by no means illiterate. Yet here we can see how her capacities have been damaged by this kind of training. The blight was not confined to a few weak candidates: indeed, it is more depressing when it comes from the more competent candidates, as from student 17:

In fact it would be true to say that from Pope to Wordsworth from the ripened Augustan age to the Romantics the social and philosophic interests of the country left a permanent mark on the style, form, content and presentation of English poetry...English poetry had progressed from Pope's 'economy of words in the best possible order' to Wordsworth's lost spirit in ethereal clouds...Pope had a pastoral piety...

In such clever hands the essay answer becomes a sequence of sentences which obviously have little meaning to the student, apparently delivered with some conviction. With the weaker candidate the answers are sometimes simply ludicrous:

There were many external influences on the thinkers of the seventeenth to eighteenth centuries. The French Revolution, & the American Wars of Independence affected their approaches to life. They began to see liberty as an ideal, & became very restless. *(Student 9)*

Such disastrous consequences of the examination system come because they are based on an essential pretence—that students are capable of much more than they are really capable of. It is the syllabus which has the prestige, not the needs of the young growing person who is to teach. Compare student 19's answer on Pope with her poem about childhood in another paper:

Pope wrote in an extremely regular style. He made use of the rhymed couplet, which in an inferior writer could have a jingling effect, but it only gave his work an added polish from which his cleverly ordered thought and wit shone.
(Student 19)

These are borrowed phrases: there is no conviction in her repetition of them. Yet she could have begun to discuss Pope relevantly, as she reveals when she says 'there is nothing of the wonder and mystery of human nature in:

> She, while her lover pants upon her breast,
> Can mark the figures on an Indian chest...'.

But compare these gropings after sophistication with the almost banal childishness of her poem *A Child in Summer*:

> Caves and sand and sea
> And boats with floats like tangerines
> Sailing out so free like me
> In summer by the sea
> I need a torch and lots of string
> And in case we lose our way
> I'll draw a plan;—
> John's got a knife;—
> And mark on treasure we may find—
> 'Oh—the teacher's there
> And I haven't done this sum...'   (Student 19)

The student is always much more immature than we tend to assume: but to accept the immaturity and work forward from there is a mark of strength. The sad little revelations of this immature self can help to lead into an understanding of Blake, or Edward Thomas, or it could be brought to feel for Pope's renderings of anguish and squalor in his lines on the death of Villiers. The inculcation of an apparent knowledge-ableness, with no basis in reading and response, simply arrests the necessary pains of development or leaves them unfostered. So, the students are writing about nothing, disengaged both from themselves and from poetry. The girl who showed herself above so inept in discussing the twelve-year-old's poem (above, p. 83) confuses *Kubla Khan* with *The Ancient Mariner*, and says of the *Solitary Reaper* that 'the girl is mourning the death of some close relative lost in the war...'(...'old unhappy far-off things, And battles long ago'...?).

Doctor Johnson's lines *On the Death of Mr Robert Levett, A practiser in Physic*, according to another girl,

show...the effects of the beginnings of the industrial revolution with the mines in his poem...He talks of 'Misery's darkest caverns known', 'the sudden blast'. All meaning the dangers of working in a mine. The times in which Johnson lived also affected his writing.

The lines referred to are:

> Condemn'd to Hope's delusive mine,
>   As on we toil from day to day,
> By sudden blasts, or slow decline,
>   Our social comforts drop away...
>
> In misery's darkest cavern known,
>   His useful care was ever nigh,
> Where hopeless anguish pour'd his groan,
>   And lonely want retir'd to die...
>
> Then with no fiery throbbing pain,
>   No cold gradations of decay,
> Death broke at once the vital chain,
>   And freed his soul the nearest way...

Perhaps we should read into the last stanza a reference to the chain-making foundries of Birmingham or the abolition of the slave trade? Or find in 'See Levett to the grave decline' a reference to pit shafts?

Whatever did some students suppose they meant, by such remarks as the following?—

Pope used many words which he thought of himself...Many of them could not be understood by the readers...the themes were usually boring and uninteresting...   (*Student 20*)

During this time, emphasis was laid on concrete imagery, serenity, collected-ness and tranquillity...   (*Student 21*)

Pope sees nature as it is. He blames man for causing animals and birds to show off, e.g. *Ah! what avail his glossy, varying dyes...*

...Mention must be made of Thomas Gray who although we think of as a transitional poet touched upon the bud that the Romantic Revival was to burst open... (*Student 22*)

This student receives 'Good, 26' for his collection of generalisations, and he was a 'good student'. Below we shall see his method of work in detail.

Here is an extract from an essay by this student on Wordsworth. The paragraph has some logic and the expression is competent. The only trouble is that none of it is true: it is composed of not very relevant half-truths, a stereotype answer, from notes:

Wordsworth however rebelled against poetic convention of the 18th century. Gray said that 'the language of the age is never the language of poetry', yet Wordsworth in his Preface to the 2nd Edition of the Lyrical Ballads said he was going to write poems using 'the real language of men.' He was going to write about simple subjects in a simple way. He objected to the slavish imitation of Augustan diction and subject matter by the transitional poets. Wordsworth, like Gray, was deeply concerned about human nature, but he does not try to lull his readers into a resigned frame of mind. He wishes to lead them to a better life through communion with nature. His poetic subjects such as Michael have experienced tragedy but have overcome it through living close to the bosom of nature. Wordsworth believes that the 'still sad music of humanity' can be cured. Thus he urges his readers to love and worship nature in all its aspects and they will lead themselves to better life. To him nature is invested with a supernatural power, in fact God.

(*Student 22*)

This is not drivel, as so many of the answers are. But it can only be called a re-hash of literary history, with some erroneous general statements about the nature of Wordsworth's work. It does not seem to strike the student, for instance, that there might be a difference between what Wordsworth said he was going to do, and what he did. It bears the marks of competent facility, but a fundamental lack of interest in the poetry itself: he is talking about such things as 'the bosom of nature' without real concern. It bears no indications of original comments based on a reading of the poems themselves, nor of acquaintance with the best criticism (e.g. F. R. Leavis's illuminating comparison of Wordsworth and Akenside in *Revaluation*). But student 22 is obviously the ideal examination candidate in such work as this because he is able to produce his roughly remembered clichés with ease.

Even at best such work—the staple of exam disciplines in Eng. Lit.—

encourages some suspension of the individual intelligence in favour of panic-stricken attention to the memorised patter. If things go wrong— well, presumably these students were awake while they were writing? Or perhaps they were desperately short of time?

The political philosophy was one of mecantilism a period then governments were liable to interfere with god's order of things...   (*Student 23*)

Nature became especially in Wordsworth a means of escape from the drudge of society...   (*Student 24*)

On the 'lines on the death of Mr Levett' we see that Johnson did not see the world through rose tinted spectacles. His lines are end stopped and the poem is full of personification...   (*Student 25*)

Pope	*Wordsworth*
Classical	Romantic
Man in society	Man as an individual
The town	The country
Poetic Diction	Common word usage
Upper classes	Common people

(*Student 18: reproduced thus in exam*)

Against this correctness of style and metre the Romantic poets grew up. At first they were few amongst many, and had little effect on the poetry of the day, but through the perseverance of Blake and Reeve the Romantic Movement began to develop...   (*Student 25*)

This student was failed. In discussion of his paper one of the college lecturers wrote:

In my course I have never used terms like 'classical' and 'romantic', and, in fact, avoid discussion of literature in terms of 'movements' and 'influences', etc. I have tried to show students that all their comments must be made according to their own inferences made when studying the poems themselves. I have in fact tried to encourage them to 'read poetry better'. This student seems to have deliberately ignored all my teaching—and to have 'mugged up' a ready-made answer to a question he 'anticipated', using a 'history of Lit.'. This answer seems to me worthless.

But nearly all the answers to this question, even among those who passed, seemed to me worthless too. (I wonder who 'Reeve' is?) The truth, as indicated by this note, seems to be that in their exams students 'regressed' to habits inculcated by G.C.E., bluffed where they could, and made it all up where they could not.

Later I shall discuss some of the good answers to questions in this exam which involved the students in local analysis of poetry. On the whole the students at this college showed themselves reasonably articulate, and obviously good work was being done.

But why do both lecturers and students seek to preserve false disciplines, and why does the college continue examinations in them? Surely everyone can see that the whole thing is a waste of time and effort?

To answer this question, as so often, we have to look at the grammar schools, and the universities. The universities say they cannot do without A level, and the grammar school staffs, pressed by governors, parents and 'employers', maintain O level—against the advice of the Schools Council—even for pupils who are going on to A level. It is a matter of habit, but also surely a mark of the deficiencies of much grammar school teaching by those who do not have sufficient aims of their own, or sufficient understanding of their work, to challenge the effects of the O level language paper? That these problems are not understood at university level either was revealed by the debate at Cambridge on the new 'Use of English' paper. The failure is in the concept of what literacy is, what reading is. Not only is it possible to pass the kind of exam I have been discussing without reading the poetry and prose—the works, at first hand. It is also possible to pass the O level language paper or the 'Use of English' paper while remaining essentially illiterate. All candidates require are certain false linguistic tricks.

It is obvious that at the college of education the students ought to unlearn these habits and disciplines—that is, if anyone in the college sees what is wrong with them. A first move, on the part of the colleges of education, should surely be to remove at once the pressure of the kind of Eng. Lit. examination I am discussing, which obliges students to waste their energy on these inefficient and inhibiting exercises. They must be encouraged to begin to learn to read, to select, to criticise, and to write— all at first hand, in their own voices. The grammar schools have often failed to make them literate in these ways. Behind this appalling tradition of preparing and scribbling more-or-less nonsense—good enough for the examiner—loom the monolithic mediocrities of the General Certificate of Education English Examinations, with all their poisonous influence on English taste, literacy, and capacity for creative interests and positive approaches to the art of the word. Even if it seems impossible for the grammar schools to break with this tyranny, because of outside social pressures, the college of education has no need to reflect it. If it does, it is simply perpetuating an approach inimical to language and literature.

I give in the next chapter an essay by student 22, a good student—a top student—as judged by this examination. The reader may judge for himself what kind of discipline this represents. Student 22 has been forced to turn himself into one of the best parrots; we have seen what the activity can be at worst. To study the consequences gives one an impression of a kind of lunacy, which is all too prevalently supposed by those working at it to represent an educational training of value. To recognise that this is so is deeply depressing to anyone who cares for English or Education.

7

HOW TO BE A GOOD PARROT

Given below, side by side, are an essay written by a student as part of his English Literature work in term, and the same essay scribbled out from memory in the examination. The essay itself is obviously concocted from various sources, more or less at third hand—the student himself almost certainly had never read Pater, or Marlowe, and certainly not Heine. The essay consists mostly of generalisations, and includes no close analysis of poetry. At one point the student even quoted Coleridge's lines about a weird phantom scene in a tropical ocean from *The Ancient Mariner*, as evidence that the Romantic Poets drew for their material on 'simple everyday life'. Though the lines speak of 'noon' he took them to be an example of writing about sunset. Obviously for him the meaning of the poetry did not come first.

The student's collection of scraps of extraneous information, and his hotch-potch of gestures at an account of 'Romanticism' (an unprofitable term at the best of times), is later reproduced parrot-fashion, from simple memorising, in the examination. The examiner, who awards it a mark of B—, writes, 'Well devised. Sound arguments and sense of planned answer.' But we have nothing that is 'planned' at all!

The essay as written in term	The essay as reproduced in the examination
Q 'In what ways was the Romantic Movement in English Literature in the late 18th and early 19th centuries a Revival?'	Q 'The Romantic Movement is sometimes described as a "revival", sometimes as a "revolt". If the first, what of? If the second, what from? Or is neither description satisfactory?'
A Romanticism was not merely limited to the period at the end of the 18th and beginning of the 19th centuries. Romanticism is, in fact, an emotional tide which has ebbed and flowed throughout literary history.	A The 'Romantic Movement' took place at the end of the 18th and the beginning of the 19th centuries. Yet Romanticism is an emotional tide which has ebbed and flowed through literary history, and it reached its

The movement reached its peak in the works of Shakespeare and Wordsworth yet it is apparent in other literature, noticeably in the works of Marlowe and Scott. Romanticism itself is the expression in terms of art of sharpened sensibilities and heightened imaginative feeling. When we speak of the Romantic period we naturally think of the 18th and early 19th centuries. Yet the word classical has been mentioned with reference to 18th century literature particularly. Pater defines the essential classical qualities as order, tranquillity and clarity. These qualities are indeed apparent in the literature of Dryden and Johnson. The essential elements of the romantic spirit according to Pater are 'curiosity and the love of beauty'. But are these, the one intellectual, the other emotional, the only essential elements of Romanticism? Surely one of the most insistent features of Romanticism is a subtle sense of mystery and this element should be added to an exhuberant intellectual curiosity and an instinct for the elemental simplicities of life.

The supreme Romantic movement was the Renaissance. It transformed not only English but European life, and it gave to English literature a greater clarity and a closer correspondence with the actualities of life. But this in time became artificial and one-sided so another movement was needed for the purposes of spiritual readjustment.

The Romantic revival was a natural development of the Renaissance and Reformation. The idea of

peak in the works of Shakespeare and Wordsworth.

Pater sees the essential characteristics of Romanticism as 'curiosity and the love of beauty'. Yet surely to these definitions should be added a subtle sense of mystery, an exhuberant intellectual curiosity, and an instinct for the elemental simplicities of life.

The supreme Romantic movement was the Renaissance. It not only transformed European life and thought, but English life and thought too. It gave our literature a greater clarity but it too became one-sided and a new movement was needed for the purposes of spiritual readjustment.

the dignity and importance of man as man, and the glories of the world of nature which we hear so much about at the end of the 18th century were ideas which had been fermenting in men's minds through all the political unrest of the 17th and 18th centuries, and were symptoms of a general ferment that had lasted on from the 15th century.

In addition the social theories of Rousseau which might sum up as 'the return to nature' materially affected doctrinaires like William Godwin and later Shelley. The battle cry of the Revolutionaries in France 'Liberty, Equality, and Fraternity' impressed itself on the young Romantics Wordsworth and Coleridge. But the general characteristics of the Romantic revival (the dignity and importance of man as man, and the glories of the world of nature) were parallel with the French Revolution, not derived from it. They were expressed in verse and fiction of Pope's time and had impressed many men's imaginations long before the overthrow of the Bastille.

The element of the sense of mysticism is expressed in the Romantic revival in philosophy, in history and in the attitudes of the Romantics towards Nature. When Romanticism touched on philosophy it favours mysticism and idealism. In history the new awakening led to the study of the last, especially mediaevalism. Many of the Romantics saw in mediaevalism a richer inspiration for the mysterious forces they felt about them. So they turned from modern

Heine has no doubt that the 'Romantic Movement' was a revival. He sees it as a reproduction of the life and thought of the Middle Ages. Romanticism broadly speaking is an extraordinary sensation of vision and feeling, and the Romantics found in mediaevalism a richer inspiration for the mysterious forces they felt about them. This may be traced particularly in the prose and architecture of the period. The Romantics were intellectually and artistically conditions towards the folk lore and legendary wealth of the Middle-Ages. Mediaevalism played such a part in the Romantic revival that Heine saw it as the reproduction of the life of the Middle Ages. Yet Pater, shrewdly, observed Mediaevalism as an accidental, not an essential characteristic of Romance.

inspired by the Middle Ages. Horace Walpole reproduced all the characteristic features of the Gothic architecture of the Middle Ages in his imitation castle at Strawberry Hill, Mrs Radcliffe reintroduced the sensationalism of the Gothic novel in her 'Castle of Ortranto'. Both of these sparked off a revival of an interest in an age which gave us our first cathedrals and churches.

Yet Pater shrewdly observes Mediaevalism as an incidental not essential characteristic of the 'Romantic Movement'. The glories of the world of nature and the dignity and importance of man as man were theories that had been fermenting in men's minds through all the political unrest of the 17th and 18th centuries and had been going on right from the 15th century. They were part of a wider movement that sought to bring us back to the bosom of nature. Rousseau to some extent pioneered the 'Romantic Movement' of the late 18th, early 19th centuries. He it was who emphasised the dignity and importance of man as man and dwelt upon the power of human love.

The Romantics were driven into strange depths of thought and feeling, beyond that of normal human experience.* But so was Marlowe in his world of moving visions, and Scott in his fervent mediaevalism. Ultimately those who had felt this power of Romance, came back to the reality, but on a higher level.

* Significantly enough some of the more interesting and true parts of his essay, such as this sentence, are forgotten by the candidate. What he remembers are the widest generalisations. Thus is revealed the process by which this kind of examination exerts an anti-educational function and damages literacy and response in favour of the cliché.

Romanticism, broadly speaking, is reality transfigured by new power of vision and feeling. Both Scott and Marlowe are realistic because of their Romanticism but Marlowe died too early to realise it perfectly.

It was the element of mysticism in mediaeval life that appealed most to the Romantics. Shakespeare loved nature for what it was, he did not see into it. Wordsworth however, found brooding and tranquillising thought at the heart of nature, and Shelley found an ardent and persuasive love. Thus Wordsworth and Shelley spiritualised nature.

There was an intellectual curiosity by the Romantics for the Middle Ages. They were artistically and intellectually inspired by Mediaevalism. This is particularly apparent in the works of Horace Walpole and Mrs Radcliffe. Walpole was genuinely interested in the Middle Ages and so he constructed his imitation castle at Strawberry Hill. The Castle of Ortranto reminded men of feudal times and created a new interest in the past. It sparked off the revival of ballad poetry, the study of mediaeval arts such as woodcarving and the greatness of an age which gave us our finest cathedrals and churches.

The intellectual curiosity of the Romantic poets re-generated English poetic style. Early dissatisfactions with 18th century convention led to a mere imitation of Spenser but Percy's 'Reliques of Ancient English Poetry', a collection of old heroic ballads, reminded men of the metrical inspiration to be found elsewhere than in Dryden and Pope. Percy's 'Reliques' were published in 1715 and almost a century later the peculiarities of these old ballads gave fresh inspiration to great poets like Coleridge and Keats.

We must bear in mind the intellectual power of Wordsworth, Shelley and Keats. It is impossible to ignore the suggestiveness of Wordsworth's poetic theory, of Shelley's transcendentalism, and Coleridge's critical insight. Shelley is clear and consistent in his philosophy of nature, and in his finest lyrics such as 'The Cloud' and 'The West Wind' there is not only a logical development, but, when the poet is so inclined a scientific accuracy. Broadly speaking, the Romantic writers transformed criticism into the art of revealing beauties. They brought their creative imagination into the interpretation of great writers.

The instinct for the elemental simplicities of life may be traced in both the poetry and prose of the time. Rousseau was the pioneer; he it was who eloquently emphasised the dignity of man as man, and dwelt upon the power of human love.

The new attitude to Nature was part of a wider movement that sought to bring us back to the bosom of nature. Thus we get the idealising of childhood by Blake and Wordsworth, and the simple nature poetry of Burns, Wordsworth and Coleridge. The Romantics realised that the sense of mystery which writers had sought in the remote past was capable of satisfaction close at hand.

We cannot ignore the intellectual curiosity of the 'Romantic' poets. We cannot ignore Wordsworth's poetic theory, Coleridge's critical insight, and Shelley's transcendentalism. Wordsworth found brooding and tranquillising thought at the heart of nature. Shelley found an ardent and persuasive love.

The instinct for the elemental simplicities of life shows itself in the idealising of childhood and nature by Blake, Coleridge, Burns, Keats and Wordsworth.

The great Romantic poets were not only inspired by the Middle Ages and Greek art but by simple everyday things. A walk over the hill, an evening sunset, the song of the nightingale, and the wind were a few of the subjects that inspired Wordsworth, Coleridge, Shelley and Keats to great achievements.

Not only were they inspired by the Middle Ages and by Greek Art but by simple everyday things such as a walk over the hills, an evening sunset, a bird, or the wind. All these subjects inspired the Romantic poets to write poems of great beauty.

Here follow in the essay examples—of which one has already been mentioned—the quoting of Coleridge's 'All in a hot and copper sky, The bloody sun at noon' to show how 'a *sunset*' can inspire a poem. But I will cut to the end of the essay:

Romanticism like every great movement had its limitations. It was essentially a school of ideas and of splendid generalisations. The Romantics did not seriously attempt to apply their ideas to the concrete problems of the day. The emphasis was on the individual rather than the whole. Romanticism in its reaction against the previous age sought to escape from modern conditions of life.

...Romanticism also too readily accepted what was primitive, wild, strange, and picturesque...

We must remember that our literature needed a new coat of paint, a vivifying and expanding influence, and this is what the writers of the Romantic revival achieved.

The Romantic Movement was therefore to some extent a revival. It was also a revolt against the poetic diction and subject matter of the Augustan and transitional poets...Wordsworth...chose simple subjects and it was his intention to make men think...'I pipe a simple song to thinking hearts'. The Romantic Movement was therefore both a revival and a revolt.

(*Student 22*)

The examiner gives the examination answer 24 marks: 'B −, well devised. Sound argument and a sense of planned answer.' Student 22 secures 75 marks for his paper, and gets 69 per cent in the English Literature examination—a top candidate. But his answer, as we can see, is 'planned' only in a very stilted way. He has merely memorised, parrot-fashion, a great many sentences from his essay written in term, sometimes even getting the commas in the right places and the emphases as 'knowing' as possible ('...as Pater *shrewdly*...'). He has obviously read few of the texts he discusses and does not really understand them, otherwise he could hardly say what he does say about Shelley's 'scientific accuracy', or Coleridge's lines. Where there is a sentence in the original essay that makes sense, he tends to forget it in the exam answer. He remembers more of the beginning of his essay than of the end, either because when he was revising the evening before the exam he was fresher at the beginning in his task of committing to memory or, more probably, because he began to flag half-way through the exam question. Whatever he is doing, he is not writing sufficiently from his own possession of literary works to be able to put together in his own words what he has learnt about the Romantic Movement, in a relevant and meaningful way. He is just unloading what he has picked up, by way of externals.

The whole activity represented by this process, by the 'best' student in an English Literature examination, is worthless—worthless as an education in 'literature' and worse than useless for a prospective teacher. This would be so even if the original work was of greater value as the delineation of a literary period.

This is not a criticism of student 22 or his teachers alone. Hundreds of thousands of pupils and students perform this valueless task every year, in the General Certificate of Education examinations, in university examinations, and other such tests. The fatuity has been exposed many times—yet such Eng. Lit. examinations cannot, for some reason, be shaken and dislodged, despite their futility. Time and energy go on being wasted, while the study of literature is thus brought to aridity. And yet, because of the awesomeness of the examination ritual, intelligent people go on believing that this all does some good, even though the contrary evidence is before their eyes while they are marking. Those colleges of education which are in the fortunate position of being able to devise their own internal examinations should break this poisoned trance by doing away with all examinations that encourage the unloading of 'stock answers' of this kind. If there are to be tests, they must be ones which foster proper disciplines of reading and writing. But a 'piece of work' is a far better substitute.

The work of this 'best' candidate reveals, when studied, that an examination of this kind can be successfully negotiated by the good parrot who can learn off paragraphs untroubled by a sense of the insincerity implicit in the procedure. Thus student 22 writes elsewhere:

The chief instrument of Swift's satire is irony. He could not help being ironical. Here was a complete ease in expression. Swift laughed at people not with them. His humour therefore is of a grim unsympathetic kind...

(*Student 22*)

The phrases are not his own, but are repeated from lecture notes, though he has enough understanding to make them sound convincing, and so the answer appears 'well planned'. That they are the vaguest half-truths does not matter—he is pressing the buttons, as it were, on the teaching machine and the stereotypes light up, 'satire', 'irony', 'laughs at not with', and so forth. When the repetition goes a little wrong in the machine we get the distortions, revealing that he does not really know what he is saying (yet even for this answer he received 'good', 26).

A man trained to repeat with facility material he does not understand is likely to be encouraged to be superficial in other directions. I feel this same student's lack of discrimination in dealing with newspapers (revealed in another essay) is not unconnected with his success in the superficial requirements of an Eng. Lit. exam.

I decided to read to the children a series of newspaper articles to show them how the journalist attracts his readers attention by using witty or amusing headlines. The object of the exercise was to ask the children to pretend that they were journalists and write one or two articles for a newspaper...

I read to them some simple items from some newspapers and emphasised particularly the use of a 'catchy' headline...I showed them how the use of alliteration here is most effective...I really think the children appreciated newspapers a little more after the exercise. (*Student 22*)

Of course, there will be many students like student 22, and they will do useful and valuable work. Compared with others, student 22 shows a sad lack of the impulse to question things, to puzzle and challenge what he is told, or what he reads. Yet, as I shall show, he could respond to good teaching, and develop a voice of his own. Where he is un-critical and insensitive it often seems to be the effect on him of a kind of teaching geared to the Eng. Lit. exam.

This same student 22 wrote a competent answer on Keats's *Ode to Autumn*, which is marked 'Sensitive appreciation B/B − : 25'. But the chief feature of this answer is that it is *not* sensitive—he approaches the texture of Keats's verse in the same mood of false 'appreciation', mechanically, as he approaches the value of 'alliteration' in headlines.

It is, again, the repetition of second-hand sources, as one can tell from the grandiose phrases:

In stanza one which is a symphony of colour Keats uses alliteration and hyphenated words to bring the reader to feel the beauty of autumn...
In the last line the alliterative 'mm' sound in 'Summer has oe'r-brimmed their clammy cells', reminds us that the hidden image of the bees has been present throughout the stanza. Thus the stanza is a whole unit in itself, just as autumn is a whole, rich season full of fruitfulness, too.
...The stanza is a symphony of sound. Keats makes full use of adjectives, nouns and verbs evoking the sound and impressing it on the reader's ear e.g. wailful choir, gnats mourn, loud bleat, hedge crickets sing, treble soft, red-breast whistles and swallows twitter...all contribute to evoke the peace and comfort which Keats' soul receives from the season...   (*Student 22*)

The writing has the air of 'this is the kind of thing they expect me to write about poetry', rather than any personal communication of perceptions of the nature of verse. There is no attempt to explore more sensitively the meaning, texture and flavour of such words as 'gathering', 'mourn', 'plump', 'treble', 'twitter'—all is dissolved in the generalities about 'symphony of sound', 'feel the beauty', 'peace and comfort'. He refers to 'satiety', but fails to discuss the vitality and uncloying exactness of Keats's language. He recognises that this element in Keats's verse is significant, since he remarks:

Keats does not seek after fact or reason except as we shall see in stanza three. But even that is only momentarily. Keats is so inspired here because he realises through nature that beauty once seen is grasped for ever. Previously he had lamented the insecurity of man's hold on the beautiful...
                                                              (*Student 22*)

Yet this is vague and imprecise. His critical writing is poor because so much of his work has been rote-learning from second-hand lecture notes and dull literary histories. He tends to write the generalisations which bring success in examinations: he is only being pragmatic—for, as we have seen, it pays to write like this. The question of the particular exactness and richness by which Keats achieves his hold on evanescent experience, and enriches our perception by his verse, becomes irrelevant to the need to do well in an exam.

Student 22 shows here and there that this need not have been so. Had he been given Leavis on Keats—or *Keats*—rather than Hough or Herford say: or had he worked in seminar discussions on the nature of Keats's poetry directly, and not at second or third hand, he could have come closer to the words. I do not mean that he should merely be encouraged to repeat the words of critics of whom I approve: but if he is excited by Leavis's perception that Keats's verse even enacts the feel-in-the-mouth of teeth sinking in the juicy apple in the line

> And bend with apples the moss'd cottage trees,

he might well, by close attention to the lines, become excited by perceptions of his own. If he did, he might be able to say something original (as children in school can) about the way bodily feelings can be conveyed, in a 'whole' way, by poetry. For instance, what can be said about the movement here?

> And sometimes like a gleaner thou dost keep
> Steady thy laden head across a brook.

The line-break here is surely used to enact the physical movement of seeking balance: it marks the exertion of steadiness against the headiness of the autumn mood, and is so part of the relish; of a sensuousness that is not allowed to cloy or ennervate. When, as in his analysis of Keats, student 22 does attempt local analysis in a stereotype answer, he sometimes fails to remember what it was the lecturer said. Being lost to know what to do, he simply makes confused gestures—when he could have said something of his own:

The use of the hyphenated word 'air-blue' seems to suggest something close yet far away. In Hardy's mind his wife's ghost is close to him yet she is far away... (*Student 22*)

For when challenged by an 'unseen' poem he does begin to use his own voice, as in the same question:

He reflects that death cannot be brought back to life...The word 'existlessness' re-emphasizes that death cannot be brought back to life. Thus Hardy believes that it was not really his wife's ghost that he heard calling him but only the breeze...Nature brings him no joy at all but merely re-emphasizes his melancholy. This is typical Hardy creed. He was constantly conscious that nature is indifferent to our aspirations and sufferings... (*Student 22*)

In such places he indicates that he has a responsive sensibility that could be trained properly to read, and to articulate the experience of reading:

The repetition of 'call to me, call to me' is most effective. It is as though it were an echo from the past; what Hardy fears is her voice coming to him like an echo out of the past. *(Student 22)*

This is the true—simple and direct—voice of the student and has behind it good teaching, keeping his attention to the poem. Yet this last answer is described by the examiner as 'disappointing'. Beside his more facile parrot-like efforts, of course, it does look disappointing, for student 22 is not riding gaily along—he is lost. In an essay in term I would be glad to see him so lost—because it is from such stumblings that the creative exploration of experience is learnt. But the examination-dominated syllabus has trained him in picking up a quantity of 'information' to unload in gushes. And so his training has allowed him—and, indeed, encouraged him—to write insincerely, and often meaninglessly, as when he writes on Swift, in the same paper:

His was an age when imagination and emotion were subordinated to wit and reason...Mankind here felt that Swift was mentally unstable they felt sorry that a man could write so bitterly against his fellow man... *(Student 22)*

Surely, leaving aside all considerations to do with passing examinations, one would rather have floundering sincerity than confident insincerity? The scribbling down of catch-phrases more or less to do with Swift involves a suspension of the concern to be careful to utter only what one really means. If we encourage such insincerity we encourage a split in attitudes to experience and the suspension of the intelligence. Take it or leave it as English departments may do, the fact is that this is the destructive effect of this kind of examination on thought and feeling—and so, ultimately, on the capacities of teachers to teach.

Let us look at a further example of parrot-work. The value of memorising and unloading is no more valuable or relevant when the content of what is memorised is adequate to the subject. Here is a young man —student 26—repeating in his training college examination an essay he wrote during the previous week. The essay was on *The Lawrence Philosophy* ('Lawrence is a Romantic'): the examination question concerned 'large themes' in poetry. The student 'uses' the question to unload a patter on Lawrence which he has all ready in his mind for a 'safe' question.

The term's essay	*The examination answer*
D. H. Lawrence is a Romantic. Romantic temperament, anti-intellectualism, love of nature, a belief in individuality—Lawrence shared all these characteristics with the Romantics.	D. H. Lawrence is no less faithfull to major themes and has close affinities to Wordsworth in his love of nature, of individualism and anti-intellectualism. He has all the qualities of a true Romantic.
In a letter to Ernest Collins Lawrence writes what appears at first to be a rather overstated indication of his basic philosophy: 'My great religion is a belief in the blood, the flesh being wizer than the intellect. We can go wrong in our minds. But what our blood feels and believes and says, is always true.'	Lawrence's major theme is of love, of human relationships in relation to a differing balance of the physical and mental.
	'My great religion is a belief in the blood, the flesh being wizer than the intellect. We can go wrong in our minds, but what our blood feels, sees and believes is always true.'
Can we separate mind from body so easily? Perhaps he follows the desires and beliefs of the body because the intellect or mind requires him to do so. There is always a danger of taking a statement out of context and not allowing for exaggeration used for emphasis. Lawrence certainly did not advocate sheer animalism or destruction of the mind. He was merely trying to restore a balance. We lean too far towards the intellectual. 'The real way of loving is to answer to one's wants. Instead of doing this, we talk about some sort of ideas.' One shudders to think of this statement in the hands of the elderly spinster. 'Does this young man think he can take the icing without any preliminaries?' But is it the	Lawrence does not advocate sheer destruction of the mind, it is an overstatement which tries to stress the importance of a correct balance between the physical and mental. We intellectualise and rationalise too much and distrust our inner feelings when these are often right. It is our etiquette rather than morals which makes us conform to the pattern of society.

voice of moralism or of etiquette? I would unhesitatingly say 'etiquette'. We all chase after our desires, these are the fundamentals; but most of us carefully conform to social etiquette. The Victorians of course, were masters of this bluff! In the Fox there is something between Henry and Madge as soon as they meet, something in the blood. 'There in the shadow of the corner she gave herself up in a warm relaxed peace, almost like sleep, accepting the spell that was upon her.' She identifies Henry with fox, which symbolised the inadequacies, the unsatisfied in her. He too feels this racing of the blood, at a glance. 'Her dark eyes made something rise in his soul with a curious elate excitement when he looked into them, an excitement he was afraid to let be seen—it was so keen and secret.' We most of us chose our 'mates' this way—a feeling of something a little deeper than desire occurs and already there is some 'communion' between two people. This is the feeling in the Blood which Lawrence tells us of. This is stronger and more basic than any mental capacities...Lawrence explains in a letter to Trigant Burrow how the mental and physical are in fact enemies in his doctrine of love: 'I am not sure if a mental relationship with a woman doesn't make it impossible to love her. To know the mind of a woman is to end in hating her. Between man and woman it's a question of understanding or love, I am almost convinced. There is a fundamental antagonism between

For example there is something between Madge and Henry (the 'Fox') almost as soon as they meet, there is something in the blood, something deeper than desire. They have instant communion at a glance, without the necessity of a word. We most of us chose our mates this way. There is something in the blood in which we can't quite explain. Lawrence elaborates this theme of love in terms of the physical and mental in a letter to Trigant Burrows:

'I am not sure if mental relationships with a woman doesn't make it impossible to love her. To know the mind of a woman is to end in hating her. It's a case of love or understanding I am sure. There is a fundamental difference between the mental and physical cognitive mode of consciousness.'

the the mental cognitive mode and the physical or sexual mode of consciousness.' Certainly successful relationships between two people are often achieved by each being blind to the other's shortcomings. To spoil the relationship would be to rationalize these failures or even try to cure them! However, to divorce understanding from love is a little hard to swallow—but I think this must depend on one's interpretation of understanding. Are not understanding and love part of the same whole?

This antagonism between the mental and physical modes of consciousness seems the major theme of Lawrence's short stories. In 'The Sun' we have a husband who is conscientious, dutyfull and respectful, and yet not loving in the emotional sense. His wife on the other hand is predominantly motivated by the physical. She gives herself to the sun, and when her husband finds her sunbathing in the nude, he is completely out of the picture, and still would be even if he stripped. His wife would have a far more suitable relationship with the Peasant than with him. A similar situation occurs in 'The Woman who rode away'. Her husband was far too devoted to work, far too idealistic and not physical enough to satisfy her. She thus made a neurotic escape but what an ironical twist—'The Indians were immune to American feminine charm.' Again in 'England my England' Egbert and Winifred had a perfectly happy relationship predominantly physical, but then along came children and 'a maternal

This antagonism between the mental and physical is a major theme in Lawrence's short stories. It might be noted that this balance between the physical and mental was discussed by Butler in *Erewhon*.

In the Sun we see two married people of two opposed temperaments. The husband is conscientious, dutiful on a mental plane, but can not satisfy his wife in the physical sense, a love she really needs. She could have a more satisfactory relationship with the peasant. She gives herself to the sun and when her husband finds her naked in the sun he is a complete misfit, even if he did attempt to take his clothes off. Similarly in 'The Woman who rode away' we find a husband who is again dutiful and conscientious but devoted to work. He can not satisfy his wife's physical needs and so she makes a neurotic escape. However what a dramatic twist—the Indians are immune to American feminine charm. Further in 'England my England' we see a couple—Egbert and Winifred—who have a more than satisfactory physical relationship until children come along. Then, a maternal sense of duty (in the mental sense) forces Egbert out of the picture.

sense of duty' was enough to force Egbert out of the picture.

In a letter to Lady Cynthia Asquith Lawrence examines the motives of national pride and war, relating to his conception of love. 'The one quality of love is that it universalizes the individual. If I love then I am extended over all people, but particularly over my own nation. And how can this be in war when the spirit is against love? The spirit of war is that I am a unit, a single entity that has no intrinsic reference to the rest.'

......

'...who touches and transmits...the life of the universe...' 'Touches' and 'transmits' are perhaps two of the most important words in Lawrence's philosophy. For if in our work we can transmit life into our work, our work will be good. It is still a case of 'Give and it shall be given unto you'. This is still the truth of life. But, sexless people transmit nothing!* 'As we live we are transmitters of life. And when we fail to transmit life, life fails to flow through us.'

Lawrence expands his ideas of love in a letter to Lady Asquith, when he defends his revulsion for War: 'The one quality of love is that it universalizes the individual. If I love, then I am extended to my neighbour and finally to my whole nation. But how can this be in war when the spirit is against love? The spirit of war is that I am a single entity with no intrinsic reference to the rest.'

......

In the poem 'We are transmitters' Lawrence stresses the importance of use having life welling in us, vitality in both the mental and physical sense. Most important, we must be active in our work, for if we put our life into our work, then our work will be good.

'As we live, we are transmitters of life. And if we fail to transmit, then life fails to flow through us. Sexless people transmit nothing.' Thus Lawrence's major theme is human relationships, their failure due to a bad mixture of the physical and mental, a failure to recognize the fundamental 'rightness' of the instinct. He claims that we rather talk about some sort of ideas, than the basic necessities of life, and thus we become unattractive and sexless. The important thing is to have life, to have physical vitality in correction proportion to mental vitality... (*Student 26*)

* Note how here the student gives this remark as his own.

I will refer later to the problem of fashionable attitudes to life, taken over by students from literature and the literary world (I became weary, in reading answers to this question, of being offered Lawrence's 'philosophy of sexual intercourse' !). But whatever the value as argument of student 26's essay (for which he was awarded B+ 'a very competent statement of Lawrence's philosophy') there is surely little merit in his capacity to reproduce it from memory, spelling mistakes included. Further effects of the examination 'discipline' may also be seen: in the term's essay there are places where the young man is thinking for himself. Some of these observations are very penetrating:

Can we separate mind from body so easily? Perhaps he follows the desires and beliefs of the body because the intellect or mind require him to do so?

This has Lawrence tied up pat—for never was there a greater contribution to that mental awareness which Lawrence despised than Lawrence's own. Again:

to divorce understanding from love is a little hard to swallow...Are not understanding and love part of the same whole?

Bravo! And this is what Lawrence himself enacts, between Birkin and Ursula, and over his own marriage in *Look! We Have Come Through!* Elsewhere the man's own insight is not so good: 'We all chase after our desires, these are the fundamentals: but most of us carefully conform to social etiquette...' This is the 'englightened' view picked up from the cynical 'realism' of the Sunday newspapers. It owes nothing to Lawrence as artist, nor to the student's own good sense.

But at such points his essay written in term shows the greatest originality, and use of his own voice. What is revealing is that he does not remember, and does not repeat, any of these more original observations in the exam room. There, his essay emerges as a stereotype trimmed of all its originality, and, as with the essay on the Romantic Revival above, consists largely of remembered extracts from some literary essay he has read. The anticipation of exam requirements encourages the unreal abstraction of Lawrence's 'philosophy'; a study of, say, *The White Stocking* or *The Captain's Doll*, with close attention to the text, would have meant much more. The kind of performance which an exam demands strips the piece of work of its growing points, and reduce's the student's interest in Lawrence to one of anxiously trying to remember a handful of *bons mots* about him, and tracts from his

'Old Moore's Almanacking'. The exam helps to reverse Lawrence's own injunction, 'Never trust the artist—trust the tale: the artist is usually a drivelling liar.' What the examination requires is the drivelling lies, rather than a closer approach to the art.

By contrast, a girl student writing her study of Wilfred Owen, on which she can take the whole year in consultation with her tutor, may allow herself, without the examination nervousness, to come close to the text, in her own voice:

Exposure demonstrates vividly Owen's dramatic use of vocabulary and his ability to select the most appropriate word for his purpose. The first four stanzas all contain brilliant examples of this, such as the phrase 'the merciless iced east winds' which as well as containing three relevant adjectives, repeats the hard, cold, 's' sound, bringing to mind the fierce quality of the wind. The sense of the word 'wearied' in the second line is carried in sound to 'Low drooping flares confuse our memory...' where the duller sounds convey tiredness. And uneasiness is apparent in the words 'sentries whisper, curious, nervous...' In the second stanza again the sounds of the words echo the sense when

> *Northward, incessantly, the flickering gunnery rumbles*
> *Far off, like a dull rumour of some other war.*

The mention of the word North, together with the play on the word 'salient' in stanza one, is almost certainly not unintentional: the reference being to the Ypres Salient—always a troubled region—and the relevance of the situation. The distance of this zone from other parts of the fighting often gave the impression that the activity there was unconnected with the rest of the war. And this feeling was a common experience among the soldiers, as David Jones similarly records it in 'In Parenthesis'. '...And far away north, if you listened carefully, was always the dull toil of the Salient—troubling—like somebody else's war.' But worse than the physical discomforts was the psychological tension and uneasiness which this poem describes. The repetition of 'but nothing happens' and the soldiers' self-questionings, tell as much as they ask about the endlessness of waiting for something to break the strain. We see that to them each day is a new enemy

> *Dawn massing in the East her melancholy army*
> *Attacks once more...* (Student 27)

Yet some still suppose the exam a better test than assessment by pieces of work like this! Whatever reservations one may have about her method of articulating her experience of reading we know this young woman *can read*: we do not know if the other two students can read fiction or not, in terms of whole response.

Now it is time to look at the attempts by students in this examination to discuss poems which they had not seen before, in their own words and using the terms at their disposal. These questions provided the test of their true ability to demonstrate the quality of their training in English.

8

BUT YOU MUSTN'T FANCY!

'Ay, ay, ay! But you mustn't fancy,' cried the gentleman.*

In Eng. Lit. examinations, students who venture away from the stock essay, and undertake first-hand analysis of poems, often suffer by being marked down. Examiners, too, become uncertain, because there are no 'right answers'. The students, unable to reproduce formulae from essays done in term and memorised, are forced to use a stumbling and uncertain voice of their own. So they put themselves at a disadvantage. Of course from the point of view of our aims in training perceptive response, these answers are often the best—often revealing fresh and penetrating approaches to meaning, however hesitatingly expressed: but they are far more difficult to mark than stereotypes. Indeed, it seems irrelevant to mark them at all, which is the whole problem.

So students are, I find, marked down more often for honesty that fails to make itself clear, than for howling rubbish. No angry remarks were scrawled against the margin for this kind of note on medieval romance:

...the courtesy in battle between knights and the apparant lack of wickedness, which is present was soon overcome; the lack of poverty...

(*Student 28*)

or this on metaphysical poetry:

Their main theme was that of physical research based on abstract reasoning; the awareness of body and mind, visionary experience and supernatural happenings...the classicists...the main ones were A. Pope, Dryden, Locke and Ben Johnson. (*Student 29*)

or this on Keats's Odes:

The Odes are intellectual and not merely emotional... (*Student 30*)

But, as we shall see, a brave attempt to explore the meaning of Blake's puzzling poem *Sunflower* is marked 'rather weak' and given 'D': a typical reaction to the hesitant sincerity itself.

* See *Hard Times* (the inspection).

Much poetry is about disturbing and painful aspects of experience, and so writing about it tends to be disturbing too. Moreover, in discussing the art of the word there are no certainties. Thus there cannot really be measurement of the kind an examination demands, nor any confidence in a confident answer. Students are therefore treated well if all they do is reproduce notes more or less accurately. If they go beyond this they reveal that all is 'dangerously subjective'—and so, unconsciously, the examiner burks.

Of course, a tutor can for his own purposes set tests of response. But he only does this to discover in reading the answers himself, whether or not the student is really applying himself to the difficult disciplines of reading and discussing his reading. The tutor can furthermore observe from such work whether he has taught—whether he has provided the student with sufficient by way of terms and methods of approach to enable him to know how to read, how to reflect on, and how to make explicit and discuss the 'precipitates from the memory' of a piece of writing. There will be other indications for his work—for instance he will see the degree to which he needs to return to a piece of literature, to insist that generalisations are tested against close examination of the text ('You say Wordsworth finds consolation in Nature—does this really apply to these lines on *The Simplon Pass*?').

But this creative use of internal tests is different from what we have in most 'pretend objective' examinations. Of these, a common feature is that if a student suddenly produces an original piece of work in spite of the examination, the examiner, trained to measure up clichés, may not see it for what it is. So when a student does show the capacity to respond in depth he often gets poor marks for it. Moreover, this may be because the outstanding piece in a sudden creative burst came from a candidate who would perhaps not have been expected to produce it. It is in such ways that examinations tend to maintain a stereotype evaluation of the human beings who take them, and this is yet another objection to them. What the examiners want is confirmation (objectivity, this!) that Miss Smith is only capable of reproducing what Miss Smith wrote in her term's essays from the examiner's notes. This confirms the examiner's estimation of her capacities. If Miss Smith suddenly runs away and writes a brilliant note on a poem with no previous help from the examiner, he may even feel unconscious envy—that she should do so well without his notes! So, he may mark her, rather than a reflection of himself, and mark her badly, for having for once been larger than his estimation of her.

Let me examine an average girl's answer on Blake's *Sick Rose*. She goes directly to the meaning, without attempting to reproduce the phrases used by more facile candidates, about 'beauty' or 'a rather dreamy effect' and so forth.

Let me, however, examine the poem briefly first myself.

The Sick Rose

O rose, thou art sick!
The invisible worm
That flies in the night,
In the howling storm,

Has found out thy bed
Of crimson joy,
And his dark secret love
Does thy life destroy.

The poem is a poem of 'experience' rather than 'innocence'. It is an expression of the mature perception which recognises that the rose, as a symbol of life, beauty, feminine ripeness and love, is doomed by time, and by the inward contradictions inherent in the flourishing of natural objects in the world. The rose may not have an actual worm which is 'within' and so invisible: but even so the worm of mutability—time and decay—belonging to the same time processes as brought the rose into being, is working within it, invisibly.

The statement is thus a metaphysical statement: since the rose is a potent symbol of beauty, femininity and sexual flowering, the poem is a statement about 'conditions of the human soul'. The natural processes are symbols of processes of inner reality. So the poem's essential theme is the inseparable complex of hate and love: the inevitable ambivalence in us of the creative and the destructive; of the frequent proximity in us of joy in conflict with the envious urge to consume. These conflict over such a manifestation of creativity and growth as the rose. The poem is a statement about the envy that arises invisibly from the hate mixed with love, to threaten annihilation. The poem thus symbolises and enacts the truth of an aspect of the unconscious being, where the urge to possession in love is a 'dark love' which contains a secret impulse to destroy the loved object by incorporation.

Because it is so organised, an achievement itself, the poem moves towards acceptance, towards coming to terms with the truth, that the deathly impulses of hate—impulses to consume the object of one's desire—are inevitably commingled with love. Our fears that love is dangerous and consuming are made tolerable by reference to the sinister (but natural) processes of creation and destruction in the natural world, which are their 'objective correlative'. The short deliberate fall of the lines enacts the acceptance, in all its gravity. The bed of crimson joy is the sexual centre, where the red blood pulses in the flesh, or is shed, by consuming love or disrupting death. The 'worm' is an emanation from the 'night' of the unconscious mind. There is an external reality which correlates in the howling savagery of the animal predator and in the indifferent howling of the wind. Though it is about the fears we all have of the danger of love, the poem yet makes the urge to consume seem a natural, inevitable and tragic feature of human existence, and of inner reality. As does the individual human life, the rose eventually dies, but the 'crimson joy' in the 'bed' has at least been experienced and enjoyed in secret, even if the consumption threatens annihilation. If we are to experience joy, we must come to terms with our dark impulse to consume that which we love: once this urge is accepted (i.e. no longer 'secret') we may be able to allow ourselves to love despite the threat of destruction from within. In love is the beauty and joy—and also the possibility of creative achievement, such as the poem, which releases compassion—such as is here expressed by the rhythm: 'Ah, rose . . .'.

Now let me turn, having said something about what I feel to be the meaning of the poem, to the work of an average student in this examination:

The 'Sick Rose' is built up on conflict. Blake is occupied with the theme of death and destruction. Death is an unidentifiable force over which man has no power.

The metaphor of the rose is useful. It expresses beauty, daintiness, colour and part of nature.

Much meaning is conveyed in this short poem. The first line is arresting and the position of 'sick' at the end gives added force and almost a feeling of convulsion. [This is obviously a slip for 'revulsion'.] A rose which is so beautiful is given the meaning of repulsion.

'The invisible worm'; this line holds various meanings. When people are dead there is the saying 'food for worms'—it is the hint of the idea that the worm is going to eat up the rose. The worm is repulsive to touch being slimy and soft and it eats decaying substance. It is something which moves silently. 'That flies in the night' gives a feeling of supernatural power, 'night' expressing darkness, unknown and fear—something which grows upon one. 'In the howling storm'—this conveys destruction, eerie noise. It is an element over which man has no control.

'Has found they bed'—a force which has sought out the bed in which she sleeps at night. Also this means death bed.

'Of crimson joy'. This is an example of oxymoron, 'crimson' referring to the colour of the rose and also refers to blood. 'Joy' suggests joy in the pleasant colour of the rose. Death, too, takes joy from destroying life.

'Dark secret love' again suggests mystery of the unknown. It gives a feeling of uneasiness. 'Love'—death has 'love' in destroying: it survives on destruction.

The last line is forceful and gives the feeling that throughout one's life death is creeping round gradually gnawing its way in to obliterate in the end.

The poem builds up throughout. This is achieved by the short staccato lines. Each line conveying a different idea and ideas are building up on one another. Each line runs into the other, is quick, and there is sense of building up to the end where the last line contrasts, being slow and full of force.

(*Student 31*)

This answer secures only a mark C −. I would myself have given it B or even A −, certainly much more than 17 marks, despite its hesitancy —for to me this is a guarantee that the girl is responding to language, feeling and thinking.

She has begun to possess the essential meaning of the poem, its pre-occupation with the complexity of the psychological impulses of joy in consuming, and with the ambiguities of love and death. She recognises the human symbolism ('a force which has sought out the bed in which one sleeps at night'), and she states the paradox of love and hate which Blake explores ('death has love in destroying: it survives on destruction'). At the unconscious level she has grasped the elements of incorporation and oral sadism underlying the poem even though she does not make these explicit. She gives by rather random free association (can one honestly do more at command under exam conditions?) her responses to the word 'worm' which show that she has *read* and taken the poem in, and its evocation of unconscious associations with death and the sexual.

The faults in her work are not hers: she has found the new-found land of Blake's word art—what she lacks is a more organised terminology and method of critical analysis. She has been taught to talk about elements of poetry being 'built up', about figures of speech being 'useful', about lines 'holding' meanings. She has, on the other hand, not been helped to talk simply about poetry and how it works—as she will have to, day after day, in the classroom. She needs to be helped to say 'This line suggests several meanings', or 'several meanings are combined in this line'. Instead of writing about 'ideas' in poems, she needs to be taught to discuss words and how they enact their meaning, by sound, ambiguity and subtle symbolism. Her approach to metaphor and other figures such as oxymoron bears the mark of too much preparation in the grammar school for mechanical answers to examination questions. But at times she transcends all this inadequate training, as for instance when she writes 'It gives a feeling of uneasiness'. This sentence is hers, it is true, and perceptive—and from a number of such sentences in her work I recognise that, haltingly as she does it, she is doing the real work of committing to paper her first-hand response to a poem.

Yet for her courage she is marked down! So, too, is the following student, who penetrated directly to the heart of Blake's *Sunflower*. Both this answer and that discussed above I find admirable—not least because they are in their sincerity very much like the best work one may expect to receive from children in school, when they are excited about a poem.

The poem tells of how the sunflower, too much exposed to the hot rays of the sun, longs for annihilation. The heat of the sun represents the trials and setbacks of this life, which can all be ended and forgotten, once 'the sweet golden clime' is reached 'where the traveller's journey is done'.

The sunflower, in its fullest bloom, represents a huge blazing disc of colour; the fulfilment of nature. The 'youth pined away with desire' and the 'pale virgin' represent unfulfilment, and yet the same fate is about to overtake all of them. The sunflower has lived life to its fullest, whilst the youth and virgin have been deprived of life's sensual pleasures, and yet they all wish to leave life, whatever its meaning has been for them.

The '*pale*' virgin and the '*pining*' youth suggest that they missed the fullness of life, and have only lived in the shadows.   (*Student 32*)

I find both these answers moving—touching on the deeper emotions, linking literature with the youthful awareness of life and its possibilities and entering bravely the penumbras of uncertainty, in our dealings with the imaginative vision. Both write directly from responsiveness as children do—and the answer above is as adequate a statement of what Blake's *Sunflower* is about as anyone in the situation could make. This student is, to one's great satisfaction, discussing *meaning*. To my stupefaction, however, I found the latter answer marked:

Rather weak. The reading of the poem has produced some interesting reflections, but the answer on the whole is weak. D plus.

This seems to me a frightened and essentially uncreative response on the part of the examiner. Or, to put it another way, the comment reveals he could not read the poem himself and was not really interested in the meaning of it.

The following remarkable piece of textual analysis of an extract from *Frost at Midnight* by student 33 received only C+ (21). Such low marks reveal that examiners do not see original 'unseen' comment of this kind as a true discipline, and do not realise how difficult it is to answer such questions well:

In the first line Coleridge uses the 's' sound in 'seasons' and 'sweet' to describe the softness and sweetness of the country, the sound itself is soft and almost beautiful in these words. In the second and third lines the 'g' sound is used by Coleridge to produce the effect of the vastness of the earth and the solidity of the land beneath the 'greenness'. In the fifth and sixth lines the sounds are a mixture of 'th', 'n' and 's'. Here the effect produced is one of the atmosphere of winter with its 'soft' snow melting on the roof-tops; it is a mixture of coldness of snow and the warmth of the sun, the vowels being both hard and soft. The 'st' sound in the eighth line, 'secret ministry of frost' suggests the sharpness and hardness of frost; and in the last line the 'q' and long 'i', sounds are suggestive of night and the soporific effect of the moon...

Is the word 'general' a suitable adjective to describe the earth? Yes, because the word allows us to imagine the earth as though we were standing away from it; we see it as we would see a globe...

'Tufts' usually refers to grass; where is the connection in this context? If we imagine these 'lumps' of snow to be green, then they would look like tufts of grass growing on the branches.

What does 'smoke' usually imply? The word 'smoke' implies something burning, something on fire, therefore we get the feeling of warmth. In this context the warmth comes from the sun. We do not shiver at the thought of snow, but feel comfortable at the thought of the warm sun.   (*Student 33*)

—Such analysis, done in the student's own words, helps him to read the poetry better, and should be encouraged. It will certainly help him to become a better teacher. He sees Coleridge's intention to convey by 'smokes' the contrast between warmth and cold, as the snow steams in the sunlight: he studies for himself the visual and tactile effect of 'tufts of snow': he notices the rendering in the sounds of the crispness of frost—this is the very territory of sensitive response to poetic art—

yet this answer is marked lower than many stock answers in which there was not one phrase of the student's own.

As I have said, the students at times have found aspects of the poetry they are discussing which enable them to give the lie to the second-hand generalisations in their lecture notes. The conventional view of Wordsworth's search for 'consolation' in nature is rebutted here by a student's perception of the cruelty and horror which Wordsworth sees in the natural overgrowth at Margaret's cottage:

The imagery is very strong, the plants are in some cases personified and they reflect the hand of nature and its cruelty and they seem to suggest a passing of time and what it brings...'The honeysuckle' crowds across the porch and yellow stonecrop blinds the lower window panes...The cruel hand of nature is stressed in the description of

> The cumbrous bindweed
> Had twined about her two small rows of pease
> And dragged them to the earth...

...the plants are beginning to destroy each other...the yellow stonecrop 'blinding' the window panes suggests a horror at what is happening...

(Student 34)

This sensitive piece of perception is described as 'interesting', but secures only C+.

Obviously the students showed good sense in avoiding questions requiring close analysis—they are not safe. In any case, few showed that they knew how to write directly about literature in this way. Student 33 (quoted above on Coleridge) is one exception—he was the only student to tackle Browning's *Meeting at Midnight*. In discussing this he draws attention to the three kinds of imagery, visual ('with only six simple and common adjectives...the picture is substantial and living'), auditory (e.g. *pushing* to describe the movement of the boat, *slushy* to suggest the dampness of the sand nearest and beneath the sea at the beach) and of smell ('it is possible for us easily to imagine the scent of the beach—the smell of seaweed, sand, and salt-clogged shells and pebbles. All this is suggested by the words 'warm sea-scented beach').

He tries to convey the success of Browning's creation of excitement here:

From the beginning the poem has had a sense of mystery, and of eager anticipation. Now, this sense reaches its zenith as Browning makes the heart pound by his simplicity and brevity of description: 'A tap at the pane, the quick sharp scratch and blue spurt of a lighted match'—an image so common yet so unobserved by people usually, its vividness therefore is pushed into our imaginations.

Now comes the denouement of the poem; a brilliant, concise description of the result of all this mystery and adventure—the atmosphere suggesting smuggling and clandestine operations. In the two final lines the whole sensation, the 'joys and fears' of the two hearts are revealed by the abrupt ending 'the two hearts beating each to each'. This image is so strong that we can almost feel our own hearts beating with the lovers.   (*Student 33*)

For this sensitive account, convincing us strongly that the writer enjoys poetry, he receives only 16 out of 33 marks! By comparison another student, student 34, who simply gives five pages of stereotype on Browning, gets 22 marks for this sort of thing:

The thoughts expressed in his poetry are ones which the reader can identify himself with, and the humour which he employs is of a type which is subtle yet effective...The reader of Browning's poetry can gain pleasure from every exercise of his creative power. It does not take a great effort in reading Browning's poetry to see what he is poking fun at or what beauty he is trying to display in his landscape.   (*Student 34*)

I am sure it would be possible to find these phrases in an essay tucked away in the student's cupboard somewhere, to be relearnt the night before the examination, unloaded and forgotten. Yet this performance secures higher marks than student 33's sensitive and sincere response to the effect of the words of the poem.

In another examination rather more students attempted analyses of Clare's *Thrush's Nest* and Hardy's *The Voice*, because they had practised discussion of these poems before the examination. Here we find evidence that close attention to poems in class can produce an examination answer which is neither cliché nor inadequate, as in this answer from student 19:

The poem has the effect of being a sad, whispering ghost of a memory echoing around the author's mind. It has the sound of a whisper, repeated, rebounding, and echoing. This is achieved by repetition, as in the first line,

          ...*how you call to me, call to me,*

but more by sounds which are very similar but not quite the same, for example, 'call to me' and 'all to me', 'Let me view you then' and 'as I knew you then', 'listlessness' and 'existlessness', 'were' and 'fair' in the first verse, which is a near-miss rhyme and also 'here' and 'near' in the third verse. This gives a real echoing effect, because an echo is never quite the same as the one before it or the original sound causing it. It is almost as if the repetition in the first line...is the firm original sound and then the echo of it rebounds throughout the framework of the succeeding verses. This puts the emphasis on the call of the woman. Hardy digresses...about her appearance but it is the call around which the poem is centred. Indeed the poem is called 'The Voice'. However, the call gets fainter towards the end of the poem as though his memory is failing him. It cannot bring her back although it can reproduce in his mind her voice and appearance. It is as though he has heard her voice in the wind. At the beginning of the poem the voice is predominant, but at the end only the wind is left in the rather chilling line,

Wind oozing thin through the thorn from norward.

The call in the last line seems distant, apart, objective. It has not got the subjective presence and fullness of the first line where 'woman much missed' is added and the emphasis is on the fact that she is calling to him. The last line perhaps echoes his feeling of hopefulness

And the woman calling.

In the first line his feelings are almost rushing to her call spontaneously. The call is leading him on and evoking images and memories, for example 'Let me view you then...'. It is almost as though she is becoming an actual physical presence and the line

Even to the original air-blue gown

with its one significant detail adds to this. However, the image seems to disintegrate after this. He is reminded of actuality by the breeze and in the last line he is left 'faltering forward'.

This line with its repeated 'f' and the wavering sound of the two words has the effect of someone taking two hesitant steps forward.

It is poignantly sad. It is almost as though the 'wet mead' is his mind and feelings at the time and the thin wind is his own stream of consciousness. It is ghostly and at the same time human and there is sadness in the lines

Saying that now you are not as you were
When you had changed from the one that was all to me...

He is trying to recapture the spontaneous feelings of his past in place of his thin feelings of the present   (*Student 19*)

This seems to me an inspired comment: certainly a completely adequate answer to an exam question. It has been given 27 marks, but I see no reason why it should not be given 33 marks out of 33—who could do better? The response and the language are entirely the student's own, the account humane, moving and perceptive. From this evidence it would seem to me that this student will make a very good teacher indeed. It is remarkable to compare this piece of work, the result of excellent teaching, with the same girl's reproduction of a stock essay discussed above (p. 89) and her unfortunate failure in the 'creative' question. There could be no more direct indication of the choice before training college English departments—students will either be brought out, by good teaching, to develop their own voices—or cramped by being given false voices. '*Kubla Khan*', she writes in her stock essay, 'is unintelligible but beautiful.' Yet it could have been interesting to see what she made of the poem. Her potentialities can be richly brought out, as they are where she has been trained to read Hardy. Her sensibility itself is liable to weaknesses, failures and immaturities, as is shown in her poem above: the wrong kind of examination-work tends to hamper or distort development rather than foster her better qualities.

A large number of students in the same exam also tackled some unseen lines of Crabbe, and, on the whole, tackled them well: here again obviously some good practice lay behind the commitment to paper. While response to the general meaning was adequate, the essential problem remains that often students do not know the fundamental terms of criticism, and so often flounder. They will, of course, need to have many critical terms at their disposal daily in school work, and it is 'building up of terms' that should occupy most of the time of English students, rather than lectures on 'influences', 'styles', 'periods', 'background' and stock essays on such general 'topics'.

Here is a girl's competent account of some lines by Crabbe:

This highly descriptive poem begins in a majestic fashion, and the word 'Lo!' apostrophied as it is, invites the reader to come and take a really close look at the details of 'the heath'. The first adjective is 'withering' and sets the tone for the whole passage of poetry. 'Withering' suggests that life is still present, but will not be for long. This idea is continued in the words 'lends the light turf'...The 'rank weeds' seem vital but are in contrast with the 'blighted rye'...The shade which the charlock throws over the young life is like a bad omen. Even the young vital shoot will not be allowed to grow strong and healthy. This is made even more clear in the next line for 'clasping tares cling round the sickly blade'...The line 'the slimy mallow waves her sickly leaf' is onomatopoeic in that the line really does sound slippery and unhealthy... (*Student 35*)

The best work done in this examination was exegesis of this kind. The worst analysis showed insufficient acquaintance with terms, as here:

The rhyme of the poem is in two line verse. The pattern is regular and the punctuation comes mainly at the end of the line giving no continuity from one line to the next. The use of the language gives the poem a soft undertone.

In the first line is the aliteration of 'w' and in the second the word 'I'. In the first half of the poem the aliteration is mainly of consonants but in the second half the aliteration of vowels is used.

The rythm is pentameter again giving a regular effect... *(Student 36)*

As I continue to explore the differences between students' work in close analysis it seems that often the habits of approach to literature in the students' writing are ingrained from earlier training rather than a fault of the college of education. Some of the responsibility for the failure of the capacity to read well and write well about literature must be surely placed with some of the schools from which the students come, and their failure to give a deeper literacy than that required to pass G.C.E. The colleges of education may be blamed for perpetuating the false disciplines of preparing stock essays, as some do. But, as we have seen, others are freeing expression and enabling some students, as here, to write sincere, 'felt' and clear criticism. The trouble is that this kind of writing is difficult to mark. This surely supports my arguments in favour of abolishing the Eng. Lit. exam.

Whether or not these students develop a creative and effective capacity to respond to literature will appear when they come to teach it in the classroom. The comments they make on a child's poem in this examination show that they have a long way to go towards these essential qualifications.

9

RIGHT ANSWERS AND
YOUNG LILAC

To practise the art of the teacher successfully one has to establish a balance, between the professional remoteness and authority of the adult practitioner, and the capacity to allow the feelings to flow. Particularly in the teaching of English do we need to keep the feelings rich and flowing, while not allowing them to slop over into sentimentality and not permitting ourselves to seek to dominate the feelings of others. This delicate process of establishing an adequate relationship with pupils in the artificial conditions of the schoolroom is one which it is impossible to describe or legislate for—or even to prepare students for except by experience. As D. H. Lawrence shows in the well-known passages about Ursula Brangwen in *The Rainbow*, the young teacher has to learn it at first hand, sink or swim. The first problem is to effect a human relationship with the children without exerting a restriction or inhibition on one's feelings—in the direction of cynicism, or other forms of disguised hostility—or on theirs. The teacher needs to recognise that he must struggle to maintain and develop positive, creative attitudes. Even with difficult, unattractive and unbalanced children, he must be there, ready to receive and in a sense to 'love'.

So, in English in the college of education, it is important to encourage student teachers to use language to explore the nature of children, of their expression and of the adult's experience with them. In examinations I have from time to time insisted on such exercises. In one we asked the students to describe in prose or verse 'something satisfying' from their first experiences of the classroom. Many of the answers revealed both that the students themselves were very immature and young, and that they met at once the essential problem of establishing the proper relationship between teacher and child. I found many of these answers pleasantly naïve and sensitive:

Very shortly I had the whole class seated before me, sitting up straight, ready to meet the new year. I surveyed them and I knew that they could see to my innermost soul. Nothing was mine any longer, I belonged to them, but in return I had a great store of young vitality of life. Hair had been brushed, faces shone, and shoes were clean, with well-pulled-up socks.

At the end of the day I felt as though something had been achieved; there was a link made between the children and myself. I looked forward eagerly to the months ahead. (*Student 37*)

> The children, not yet awake
> Sat drowsily in the morning sun,
> Warmth in the classroom
> Frost without.
> Some yawn
> Others fidget
> But mainly they listen,
> Quiet but attentive...
> Gradually,
> They awake and begin to take note,
> Smile at the familiar,
> And begin to ask questions,
> Quickly moving from the story
> To their own lives—
> Aunts and cousins,
> Sisters, friends.
> Noise mounts,
> Then falls as the story continues.
> Happy are they and happy the teacher. (*Student 38*)

Such work, however simple, in the imaginative contemplation of one's experience, to seek insight, is most important for the teacher in training, and happily much of this kind of creativity is now encouraged. Response to language, the stimulation of creativity and the creative art of teaching itself—these demand a high degree of insight, and of sincere recognition of one's own limitations, one's own weaknesses, and one's own degree of civilisation. But, contrary to the implications of the information-bound syllabuses of training colleges in general, these come best by creative activities of all kinds, including teaching itself.

In teaching literature to undergraduates, to adult students, to teachers at refresher courses, or student teachers, one finds again and again that the first part of the work is to break down resistances which the 'reasoning' part of the mind puts in the way of the full response of the sensibility of the 'whole being'.

It may be that the kind of training students receive in the grammar school—preparation for 'Eng. Lit.' examinations—actually manifests in this way against their response to imaginative language in English. The assumption of a false sophistication from literary history books, the development of a 'knowing' patter about literary texts, the training in formal answers—listing the 'merits' while (of course) paying attention to the 'demerits'—all this tends towards facility and a tendency to be 'ashamed' of the 'extravagances' of the creative mind. The mind, trained in 'rigorous disciplines', ceases to be 'open' and 'relaxed'. The student ceases to be able to allow doubts, uncertainties, disturbance in his own inner world, and so ceases to respond to literature in a creative way. He is defended against it by an intellectual approach—he no longer feels, no longer knows what he feels, and is no longer capable of being moved, or opened to fresh experience. He 'appreciates' rather than speaks of his true responses. To be stirred, puzzled, moved, upset is to experience a range of reactions—trivial to fantastic—which, if he experiences them in the examination room, will lead to hesitancies and callownesses that will penalise him. Those who defend themselves against all the disturbances by which literature has bearing on life (such as student 22 above) do well, while those who are really responding to the art may be marked down. A 'planned' answer comes to mean a well-remembered one, in a mind closed to the momentary and transient, and shut to 'ideas that rush in pell mell'.

To abolish examinations and to substitute more creative work for the mugging up of information about Eng. Lit. can therefore be recommended for the most practical reasons. For when the teacher reaches the classroom it is his degree of insight and fresh creative interest in children and children's work, and his relationship with them through their work, that will count, in terms of the effectiveness of his teaching. A teacher who has written at some time or other honest creative work about his inmost self, and who has often been truly moved by word-art, will be able to respond to children's own creative writing, and will be able to select from literature those works of finer minds which can add to and develop the child's work, which work to praise or select for comment, and he will know what to say to the child.

Consider, for instance, how much the writing of a poem such as the following meant to a student teacher. I find it moving, because of its sincerity—a degree of insight, in humility, into the immaturity and childishness of the student himself which must have been painful, and which can also be a source of strength. A student who is able to accept that he is not really so far from childhood (note the title) is not likely to deceive himself by pretentiousness, and will be able to go out from himself in easy sympathy with his pupils:

Last Summer, When I was Young

I, seventeen, and yet a child
Finished summer examinations
and sheep-like followed
'The lads'
to a back-street local,
Where teachers had never been known.
I, seventeen, sheep-like
Drank pint-for-pint with them
But gradually it became more-and-more,
More-and-more an effort of will
To remain sane and sober.
At ten, I paced with studied care
between mobile tables and fluid people,
tensed in the shock of night air,
said 'see 'yer' and turned into a back-street.

It was grey light,
I could relax in the deserted street,
Need not keep up the pretence
Of 'being able to take it' no longer
And a grey pavement whirling,
Brown-brick whirling, headache whirling world
Came to rest on a sliding floor-wards
Cold green lamppost.

I felt a sticky hand
On my burning brow and,
Looking up, saw a little girl of about seven,
with dirty, ill-clad legs and clothes torn
and grimy toffee and lank drab hair
and a face as old as time
in ravaged coarse sensuality
with wide experienced eyes and a thick vulgar mouth.

The sticky hand again touched my brow
and I dreamed of my brow melting her toffee
and making her cry on my shoulder
But it could not, I knew,
And I cried in the gutter. (*Student 15*)

The poem makes us uncomfortable: it 'goes wrong' at about nine lines from the end. But it goes wrong because the adolescent who has tried to pretend he is more than a child suddenly comes to fear regression, and struggles to thrust the child away from him. Yet sentimentality and self-pity overwhelm him: he fails to place this, and his self-disgust. But even in its failure the poem marks an effort to come to terms with the weaknesses of immaturity. Having been brave to fail, in 'giving' such a poem to his tutor, such a student is showing himself brave enough to accept his own weaknesses. So he will come, as a teacher, to children's poetry with a sympathetic sense of what they, in the same way, are doing. A student who writes a poem about 'knowledge', student 33 below, has become able to see in the effort of writing it that knowledge at its fullest is a contest with experience as a whole, not the acquisition of information (it was student 33 who wrote the very fresh pieces on Coleridge and Browning above, pp. 120 and 121–2).

Knowledge

Glistening, harmless mind, with large blue eyes,
so new to life, that it lifts
What often people grime with foot-prints.
Take up the book, look up to sun and smoke,
and feel the heat of land link with friendly sky.

Let the sea be angry, tremble with rage,
but turn your back and love the leaves
quickly, before the waves catch up.

I can see you sitting;
do not rise and walk; those mountains, there,
only bring the sky to thunder at the land
and crush all the people,
and books roll down the mountains to the sea.
Sit here while the sun still nurtures.   (*Student 33*)

This student's pencilled notes for the poem are interesting: he writes, 'Knowledge, not idle. Being absorbed, kindling hearts. A child's mind being full to the brim of life, to take in knowledge.' The poem expresses an adolescent sense of the unconscious danger of 'knowledge' (cf. 'after such knowledge, what forgiveness?'), yet an awareness that only by absorbing aspects of the nature of experience can a child live and grow, be nurtured. Yet the writer has a strong sense that the 'nurture' must be whole, a whole possession of experience, warm, and

vital, 'kindling hearts'—enlarging sympathy: it seems to me an interesting poem, a truly metaphorical poem, about the nature of
learning.

It would have been interesting to have this student's comment on the
Lilac Tree poem below. This poem is a valuable 'artefact' of childhood,
useful to test students' capacities to respond to child art.

> The lilac tree stood over the gate
> Its young leafs moved in the breeze
> The little green flowers not probly out
> Heafely ladan it sways this way and that
> Soon my little lilac tree we'll be out
> Each day it gets whiter and whiter brighter and brighter
> Very soon it's like a crown.
> A crown worn by an angle
> An angle in white the best to be seen. (*Florence, 12*)

A teacher needs to be able to recognise the beauty in such a child's
poem. This is not so easy as it sounds—such naïve tenderness embar-
rasses many who would therefore be likely to miss it, by not being able
to *receive* it. He should be able to see that the poem's sincerity is guaran-
teed by its simple, 'meant' rhythm, and the absence of all verbal
'tricks' or attempts to strain at expression. A good English teacher
should be able to take in the effort such a poem 'costs', in terms of
questing vision, in the child's psychic life.

The teacher who has learnt to respond to the true voice of poetry
will register an inward satisfaction that this is 'the real thing'—one can
tell by the way the rhythm of the speaking voice comes alive in the
revealing, 'Soon my little lilac tree we'll...' (the use of Mummy's
reassuring manner). The child is registering the 'inscape' of feeling.
All a teacher can say to Florence is that she has written a lovely poem,
take steps to see that her work has the fullest possible audience, and see
what poems from literature could be used to follow up (Lawrence's
'Flower' poems? Blake? Edward Thomas?). A student teacher of
English, certainly for the secondary modern or primary school, must
be trained in such a positive reaction, by his experience of literature and
other works of the creative imagination.

The reactions of the students when I have set this poem in examinations have often been disastrously anti-creative. As we have seen,
student 11 considered it needed better punctuation, in order to
'give it more sense'. Florence's poem is *full* of meaning, one protests:

punctuation and spelling would make no difference to its quality as art. The impulse to 'improve' goes with the attitude to literature revealed in the same student's answer on Swift—she reduced his work to cliché (e.g. by saying that his pretend-societies were 'better' than English society—which at the most obvious level is simply not true). Then she set out to do the same to this beautiful poem by a child, on the assumption that she *knows*—that she *could* know—what the child 'intended': there was something in the child's mind (a real poem) which only partially 'came out':

I would suggest to Florence that the first line should read 'The lilac tree grew beyond the gate' or something similar to this... (*Student 11*)

The phrase '*over* the gate' surely enacts the poised moment of waiting to flower and emerge—through the gate—into womanhood, and the spring of youth, a moment of pausing and of poise? But student 11 goes on:

I would show her the difference in meaning between what she had written and what she had *meant to convey*. Perhaps she could suggest an alternative first line herself, which I would prefare in any case, in order to keep the poem mainly all Florence's own work.

When referring to the 'little green flowers not properly out' I think that Florence means the buds. In the fourth line she tells us that there are many buds on the tree and that it is heavily laden; then goes on to say that soon the tree will burst into flowers, becoming whiter as more flowers open. Florence sees the white flowers as being an angel's crown which is a good comparison. However, I think that in the last line she does not achieve continuity in rhyme or rhythm, so I would try to help her over this difficulty after first asking her for her suggestions. When the poem had been improved to Florence's liking, I would suggest that she copied it out in her best handwriting so that it might be pinned on the notice board at the back of the classroom. She might also illustrate it. (*Student 11*)

Here the newer techniques of creative encouragement are at odds with the uncreative academic attitude, destructive and strangely envious. This student, having never been encouraged to respond to literature as creative meaning, cannot allow the possibility of deep poetic meaning being *there* so marvellously in the little poem with all its mis-spellings and lack of punctuation. Student 11 *must* interfere to 'get it right': there is a 'right' answer, a 'right' poem. Her literary training, bound in the pretensions of 'exam' stereos, reinforces the anxiety which the exam candidate feels at being presented with anything unseen and new, anything which makes a creative impact requiring adjustment—it would hold her up. She would be flummoxed by anything which has no 'right answer'. There has (after all) been no time to waste on reading Swift, with all his disturbances. So, she reveals a resistance to creative expression, and has no sense of how to encourage it. She can not *receive*: she can not generously say 'This is good'. Examinations imply that there can be a degree of security, of 'right answer', such as there can never really be in dealing with art. (If a pupil fails to 'appreciate' a poem set for appreciation, because she does not like it—what then?) All critical conclusions are open-ended assertions: the questions in life are more important than the answers, but not in examinations! Thus does the Eng. Lit. exam operate against a true response to literature—whether professional writing or the creative writing, with its quality as literature, of the child.

So, in reading the child's poem, students tended to reproduce the niggling negative approach of their own teachers in the grammar school. They tended to look for faults and to exert a predatory authority of external 'correctness' in 'expression'. No one saw that the poem was essentially about Florence herself—about a condition of the human soul ('flowering Florence'!). Most regarded it as a kind of child's poetic nature note, showing accurate observation of 'nature' and so forth.

But the prevalent approach was one which, if it was used in practice, would very quickly put the children on the defensive, and inhibit creativity at once:

This is a pleasing poem to read but it has many weaknesses, its main one being the lack of punctuation. It is a descriptive poem and clerely shows that the child has observed the lilac tree before writing about it. It is a record, *there is no feeling there* [my italics] but in its simplicity it has a certain appeal...
(*Student 39*)

I think that this poem is very good for a child, but that a girl of twelve ought to have some idea of the significance of punctuation... (*Student 21*)

The teacher would have to help Florence with spelling and punctuation. Florence has learnt about the apostrophe but as yet does not know how to use it properly. This would have to be taught by the teacher.

Florence has not learnt properly the use of singular and plural. In her poem she puts 'leafs' and therefore the teacher would have to instruct her about this.
(*Student 40*)

There are several grammatical mistakes within the poem and this would be an opportunity to improve the girl's knowledge of English language by pointing out her errors. She uses the plural of leaf wrongly here using leafs instead of leaves. She also misspells the words properly, heavily, laden, all these mistakes seem to stem from the incorrect or hurried pronunciation of them in ordinary conversation. If the child is therefore encouraged to pronounce these words and others like them clearly and articulately the correct spelling of them will become easier. The girl has become confused with the word 'will' and the spelling of angel will also be recognised by pointing out what she has written really means. *(Student 41)*

The most important thing I would say to Florence is for her to watch her spelling and to try, where possible, to use some punctuation. *(Student 42)*

The first thing that strikes me about the poem is the spelling. My first thoughts would be that a girl of 12 who has the ability to write a poem of this nature would have the ability to spell better. *(Student 43)*

To the child I would give praise for the idea and content of the poem but I would them go on to point out the puntuation mistakes, grammatical errors and spelling errors. I would point out that stood is grammatically incorrect and even standing gives the wrong impression. I would suggest that the child should use the word 'hanging'. I would ask the child what word she ment by 'probly' and try to get her to spell the word correctly. This I would do with all the following words which she spelt incorrectly. The words being 'Heafely' 'we'll' 'angle'. *(Student 44)*

No doubt the students are so obsessed by their own spelling difficulties that it is natural for them to be so preoccupied. But there is here also an impertinent, negative function at work—an unwillingness to 'leave it alone',* which emerges possibly from a deep resistance to creativity, bordering on envy. They were unable to accept that a child can write so beautifully, and so deeply. The reason they do this, with such destructive anxiety, is that they are working in the ethos of an examination which imposes a false security and a bogus decorum. In the classroom they would become much more tolerant and positive.

Here one may perhaps quote Melanie Klein's advice to the child psychoanalyst, advice which the creative teacher can well take in for his guidance:

One of the many interesting and surprising experiences of the beginner [in psychotherapy]...is to find in even very young children a capacity for insight which is often far greater than that of adults. To some extent this is explained by the fact that the connections between conscious and unconscious are closer in young children than in adults, and that infantile repressions

* See the chapter 'Leaving it Alone' in *The Experience of Poetry in School*, ed. V. V. Brown (Oxford, 1953).

are less powerful. I also believe that the infant's intellectual capacities are often underrated and that in fact he understands more than he is credited with.*

This is the kind of 'capacity to receive' in which student teachers need to be encouraged: examinations manifest against it, as their effect is to limit expression to its most mundane and mediocre.

Here we see a further way in which exams in Eng. Lit. are inimical to teacher training. For the criticisms these candidates make belong to the world of exam answers, not to creative response. In the classroom the children's faces alone would inhibit such negative approaches. Indeed, these are often mistaken: in 'The lilac tree stood over the gate' the word 'stood' is *not* ungrammatical. It may be unusual, metaphorical: it strikes one as fresh, by personifying the tree: in fact, it is just what we want. We see a lilac tree which immediately becomes a human figure standing over a gate, as if waiting for someone to come home (like a father or mother), perhaps waiting to open the gate by which we are to walk out into the world. This figure evokes the two metaphors underlying the poem, of the child aware of herself as a flowering shade coming to season, soon to open the gate, burst into leaf. The other of her wanting, by taking on her mother's voice, to assure herself that 'everything will be all right'—that she will come to flower, that she will wear the angel's crown. The angel's crown is that of womanhood, and she can only conceive of womanhood in close contact with father and mother: hence the figure standing over the gate is her blossoming, emerging, soon-to-be-crowned self; yet at the same time the adult parent, waiting for the child to come home or seeing her off, with whom she identifies herself.

To seek to point out that 'stood' is 'ungrammatical' comes from an unconscious impulse to operate against this very subtle art of a child's simple poem. It is but a step from an excessive concern about spelling (as if spelling were the mark of a capacity with words, of 'education'), to the impulse to 'improve' the poem, by making alterations. To some students the pretty achievement could not be left: it must be brought into line with the poem Florence *meant* to write, which, they impertinently assume, they would know if they saw it. This attitude comes from a profoundly false attitude to creative writing, that it is the expression of thoughts, ideas, concepts which already exist, and need only to be 'clothed with words'. Whereas the concepts, the 'ideas' do not exist before the poem, only come into existence as it is being written, and

---

* Melanie Klein, *New Directions in Psycho-analysis* (Tavistock, 1955), p. 13.

only exist in the poem as it is written. To alter the poem is to remove elements of its meaning—to alter the line

Soon my little lilac tree we'll be out

to, say,

Soon you will come into flower, lilac tree

would be to destroy that rhythm of 'Mummy's voice, speaking reassuringly'. Such interference (which is, of course, often found in schools) would destroy the timid, childish intimacy of address to natural objects, destroy the point of the symbolic identification, and so undermine the vibrancy of feeling.

In the grammar school it often seems to be believed that a child must never be given full marks—but must always be told it 'must do better next time'. The *faults* of any composition must be pointed out first. It is always A −, never A + ! There seems to me little justification in this attitude, even with intelligent children. Children respond much more enthusiastically to encouragement, within a disciplined context of work and application. Where excuses are made for this negative approach we may suspect that rationalisation of envy is at work.

Here the results are seen. These students are over-preoccupied with externals and mechanics. A teacher primarily concerned with true literacy is confident he can continue to seek to clear up such typographical matters in regular periods of drill. But the fluency must be there first: and the first aim is to 'receive' this. These students go on to offer 'improvements' which would destroy meaning, and which by their implications and effect might well make it harder for a child to achieve literacy:

When talking to Florence about her poem my main concern would be in showing her how she could improve it. The first point being punctuation. Let her see that it cannot be read easily unless one is sure where to pause. Help to see that a pause after 'gate' and a full stop after 'out' would help and then see if she can work out the other punctuation, giving her help when required. (*Student 39*)

Florence, for my purposes, has made the need for a pause after 'gate' quite clear by her line-break.

Some of the words she uses are rather ordinary, perhaps she could think of words to replace them, i.e. 'moved in the breeze' is there a word which will suggest the kind of movement—quivered perhaps.

136

'Moved' comes from the intensity of the child's realisation of the lilac tree as a figure—it moves like a human being, showing that it is alive, sentient.

Point out the misspelling of 'heafely' 'ladan'. Ask her if she means 'will be out' if not ask her to explain the meaning of 'we'll'.

I would certainly tell her that I liked her poem and indeed I am sure it would be very good with the few alterations that I would help her to make... I do not really like 'whiter and whiter brighter and brighter'. It sounds very much like an advert, and I would see if perhaps she could think of different words or expression of the words for that line. It is by no means a piece of outstanding work and it would appear that she had very little experience of the writing of poetry, this of course is not her fault. If she was given that experience I think she would be able to produce some most acceptable work.

(*Student 40*)

One's only reaction to such predatory attitudes is 'damned presumption!'. Florence would certainly dry up at once—and would be quite justified. (This is not to say that there would not be plenty of 'practical English' to do with her in other ways.)

The primary thing to do, in my opinion, is to correct the spelling first of all and insert a minimum of punctuation to give the poem more sense than it already possesses in its present form... [*Has this student not read the last chapter of 'Ulysses'?*]

We may also recall that student 11 said, 'When the poem has been improved to Florence's liking...'

I would suggest to Florence that the first line should read 'The lilac tree grew beyond the gate,' or something similar to this. I would show her the difference in meaning between *what she had written and what she meant to convey*. [My italics.] Perhaps she could suggest an alternative first line herself, which I would prefare in any case, in order to keep the poem *mainly* all Florence's work...

(*Student 11*)

What would be the effect of such responses in the classroom on the child? She would soon become altogether sick of the thing, and regret that she had ever exposed her feelings to such insensitive teachers, surely?

The better answers revealed students who perceived that the lilac tree possessed some special symbolic significance for the child, but because of their lack of training in close attention to meaning did not penetrate to its metaphorical quality. Had they read some good literary criticism—or had they ever had a real seminar on poetry—surely they could not have missed it?

The poem is a pleasant one. The child has obviously looked at this lilac tree and she has admired it so much that it has come to assume a particular significance for her. It is a simple poem and yet it achieves a forcefulness through its simplicity. This is the best way that a child can look at natural things and develop a natural affection for them. The girl obviously feels for the tree:

Soon my little lilac tree we'll be out.

Seeing this 'humanisation', though she does not express it, this girl student is able to bring herself to the edge of a creative attitude:

The child should be congratulated on her poem and be encouraged to write exactly what she feels.   (*Student 41*)

Elsewhere I quote this student's qualifications, about spelling and punctuation—and she obviously cannot easily shrug off her exam-bound anxiety. But the above is commendably positive.

The best answer, if a little pedestrian, reveals attitudes and approaches which one would endorse: it was a relief to find such a positive answer:

Florence expresses her obvious love of nature and colour in this short poem. It is plain to see how delighted she is to be able to watch the daily progress of the lilac tree. Her appreciation of nature, and also of the colouring of the flowers are very developed, and her description:

Each day it gets whiter and whiter brighter and brighter

explains to me very clearly the opening of the flowers. Her choice of words, and the repetition, make me share some of her excitement in this 'marvel of nature'. Florence has observed the tree—all its movements

it sways this way and that

its structure—'heavily laden', the leaves and the flowers, very closely, and her accuracy of description enhances the poem.

The faults of the poem, mainly punctuation and spelling, only add charm in my opinion, and make it seem more personal to this child of twelve years old. If the structure, punctuation, and spelling had all been perfect, I think there would have been something 'artificial' about the poem, as though it were not wholly the child's own work. I think the full stop she has placed after 'crown' is very effective. When reading the poem there is no punctuation up to this point, and this makes one feel the growing excitement of the child, and then after the word 'crown' there is a pause.

A crown worn by an angel
An angel in white the best to be seen.

These two lines are almost separated from the rest of the poem as though this thought is precious to her, and it is a most effective description.

The poem is a description of a lilac tree which hangs over a gate, and is moved gently in the breeze. The child is watching the tree and anticipating the day when it will be full of white flowers, but as yet the young leaves are still green and partly curled. She can see the tree is heavily laden with flowers, and is thinking how nature opens the flowers from their buds. Florence is to be greatly praised for this work, it is certainly an achievement. Her love of nature is very clear and she expresses herself very well. I would point out the faults in the poem, but certainly not decry it because of its faults. I should find out is the tree in Florence's own garden, or if it is one she passes each day...Her choice of words is very clear and descriptive because she is obviously talking of something she loves...   (*Student 45*)

One might add that Florence is talking of something she loves because she is seeking to love herself: her poem is a simple poem of wish-fulfilment and engaged in that total egocentricity which is a condition of all our dreams, as Freud points out. Everything in a dream is a part of us, or an aspect of our identity split up. Florence is dreaming of the self who wants to grow through adolescence and who needs to find a way to love herself, during the ungainly changes (psychic as well as physical) of this time. But yet, because this is expressed metaphorically, because it has its own creative energy, as an 'ikon' of the quest for significance and beauty, it is also beautiful in itself, as an artefact, and has more than individual significance. The other children can share it, and it will help them with their self-awareness, their needs to develop a positive and beautiful sense of their own 'structure', and so actually help them through the 'heavily laden' transformations of adolescence, to-wards the joy of self-realisation.

To nourish such work in children requires a thorough training in the true disciplines of response to the creative word, and in positive atten-tion to the relationship between the organised word, the sensibility, and living experience. While there is evidence that much good work is being done in these directions, the answers given here show how short student teachers may fall of a positive approach. The only way to de-velop such an approach is by a great deal of experience of creativity and free fluency, and of the open response to the art of the word. To enable them to have this rich experience the syllabus should free themselves from those false and restrictive disciplines of 'Eng. Lit.' imposed on them by exams which not only waste time and energy, but actually impede the genuine development and release of creative potentialities.

IO

ABOLISH IT ALTOGETHER

My conclusion as an external examiner for training colleges has always been that Eng. Lit. examinations neither tested nor encouraged anything relevant to the needs of the English teacher, either as educated young person, or as a professional worker who is destined to grapple with literacy in the classroom. They merely preserved habits of approach and concepts of 'Eng. Lit.' which were academic, dead, and irrelevant. These approaches had been carried over from the grammar school, and the work required for G.C.E. and university entrance.

Anyone who has tried to improve one (as our working party of East Anglian teachers did, when we compiled *English in the C.S.E.*) inevitably finds that an 'examination' is itself a cultural artefact of very solid fossilisation. We all have an idea of what an examination is, and we fall into extraordinary habits of reproducing the traditional artefact (because we were drilled for it ourselves) while failing to question the nature and purpose of the whole manifestation. Because of the loyalty to the exam inculcated all through our schooldays we found it difficult to look at the phenomenon from a detached point of view.

The exam experience is rooted in us, and the machinery is embedded in the system. We could never allow ourselves to feel that the exam did not matter (and when we were given a 'good degree' allowance we saw it turned into bread and butter). Exams are naturally defended with energy—sometimes fanatical—by the interests which conduct them—by the boards, the universities, by teachers, by the habits of teaching in the grammar school, and by textbooks. Secretaries of boards will send you pamphlets about the tons of scripts wheeled in every year, complete with photographs of postmen and bags of papers. It all seems, of itself, because it is so huge, necessarily meaningful. Yet in English work the grammar schools have now reached a situation in which no one seems able to conceive of alternatives, or ways of dislodging their tyranny over the syllabus. In the grammar school the demands of G.C.E. examinations continue to exert a pressure that tends to drive out the humanities, reduce the time given to arts, to imaginitive work, to music and domestic subjects. It restricts those freer activities which help develop the young personality as a whole—and with them the grounds of literacy. The results are well known at the university, in aspects of the higher illiteracy. The whole effect may be indicated by quoting the remark of a teacher to a friend's son: 'You are not to think like this before O level.'

There would seem to be little justification for extending the tyranny of the examination over areas of education not yet subject to exams, though, as we know, this is what is happening in the secondary modern school. But this will go on happening while examinations are given such over-emphasis in teacher training. Having experienced them all his life, the average student teacher will unconsciously assume their importance even if outwardly he denounces them: in his work later he will feel insecure unless he has the apparent criteria of 'objective' judgement by which to measure prowess and achievement. We shall find it hard ever to shift exams unless the colleges of education get rid of them. There are new ways of assessment which are being developed: as yet not enough teachers are clamouring for these (as, for example, to be allowed to present their own syllabuses for C.S.E.). The reason may well be that they have been conditioned in exam habits at colleges of education—and this means conditioned in false disciplines.

What students should be given, surely, in colleges of education, is the experience of building up their own sense of aims in English—of what we teach English *for*. (One can do well in exams without ever asking oneself this: the exam seems purpose enough.) And their experience should be of the alternatives—of continuous assessment (and its difficulties), of 'the piece of work'—and of the gradual inward discovery of that self-discipline derived from a felt need to discover, with the help of words, some sense of the meaning of life, and how to explore it. Every young person has such an impulse—as children have: education can harness it. An exam too often becomes a substitute, and halter, with a quite different effect, as we have seen.

MEETING IN THE WORD

A different perspective of approach to English

II

O HAPPY LIVING THINGS

The needs of adult teacher and child pupil may be stated quite simply:
they meet in *the word*. The essential process of teaching English is that of
a concern with whole meaning. It may be represented by the following.
The teacher is reading a poem to a class:

> The moving Moon went up the sky,
> And nowhere did abide:
> Softly she was going up,
> And a star or two beside—
>
> Her beams bemocked the sultry main,
> Like April hoar-frost spread;
> But where the ship's huge shadow lay,
> The charmèd water burnt alway
> A still and awful red.

A child, interrupting the progress of the poem, asks what 'bemocked'
means. What is happening? And what happens at this point?

I could spend the rest of this book describing what is happening, what
is going to happen at this point, and what comes afterwards. How is it,
then, that I can say this is a simple and direct matter?

It is simple and direct, in that everything depends on the answer to the
question, 'What does "bemocked" mean, Sir?'. The English teacher
teaches by stimulating, and answering that question—and following that
answer with a great many others. But he starts from the words on the
page, of a poem, which is a work of art. And his teaching of that poem
will be controlled by his awareness of the need to bring the children to
their best possible possession of those words, in all their flavour and
complexity. Only if he does this can they possess Coleridge's poem.
There is no substitute. A parallel, not quite exact, is with music—unless
the children actually hear the notes of a Piano Sonata by Beethoven,
there is nothing for them to discuss, where 'music' is concerned. If
they do not hear the music and respond to it they are not being taught
music.

Here, unless they possess the meaning of such words as 'moving', 'abide', 'Softly', 'bemocked', 'sultry', 'hoar-frost', 'spread', 'char-mèd', 'burnt', 'still', 'awful', they have not 'taken' the poem. Unless they respond to the words they are not being taught English, for English is not in anything else. The difficulty, of course, is that to possess the meaning the children do not even need to consciously understand them, and certainly do not need to 'explain the meaning' explicitly (as the teacher may be able to) or place words in linguistic categories: as we know well enough, exegesis may not be an indication of adequate response. But one often finds students who will run away into a discussion of such a poem before they have assured themselves that they have *read it*, by taking in fully the meaning and flavour of each word.

Poetry, because it is an art, demands flux and readjustment in our responses—sometimes painful changes, sometimes depression, certainly perceptions which disturb our psychic comfort, even if the outcome, as sometimes with being made depressed, is 'useful'. So, as I. A. Richards demonstrated long ago, in *Practical Criticism*, even intelligent people are not good at taking in the meaning of a poem, and falsify the meaning sometimes to an extraordinary extent.

The tendency can be studied in detail in those chapters of I. A. Richards's *Practical Criticism* in which are given various students' responses to poems. The most revealing are those poems which have the deepest emotional content. Perhaps the best example is Poem VIII, Lawrence's *Piano*:

Softly, in the dusk, a woman is singing to me;
Taking me back down the vista of years, till I see
A child sitting under the piano, in the boom of the tingling strings
And pressing the small, poised feet of a mother who smiles as she
    sings...

The deep current of nostalgia in this poem inevitably provokes anger, because this is a disturbing area of our being. We both reject our nostalgia for infancy, and yet are all also deeply impelled to mourn the loss of childhood. To come to terms with these two contrary impulses is a problem of adult resignation to the conditions of life—to time, to the inevitable loss of the past, and to the relinquishment of past happinesses. Lawrence is simply being honest, about this area of deep emotion—and he is afraid neither of the emotions, nor of 'placing' them. When we read his poem, we have not yet, perhaps, been able, either to recognise this source of distress, nor have we begun to go through the stage of coming to terms with nostalgia such as he records. To possess the poem is to have to work on our subjective reality. So we may react thus, as one of Professor Richards's students did:

After about three readings decide *I don't like this*. It makes me angry...I feel myself responding to it and don't like responding. I think I *feel hypnotized by the long boomy lines*. But the noise when I stop myself being hypnotised seems disproportionate to what's being said. *A lot of emotion is being stirred up* about nothing much. *The writer seems to love feeling sobby about his pure spotless childhood* and to enjoy thinking of himself *as a world-worn wretch*! There's too much about 'insidiousness' and 'appassionata' for me...I expect I am too irritated for this criticism to have much value...

As I. A. Richards points out, much of this person's response was 'added' to the poem—'the long boomy lines', the poet's 'pure spotless childhood', the 'world-worn wretch' and a later scornful reference to 'passionate manhood' are simply *not in the poem*. Besides, references by Lawrence to the 'insidiousness' of the 'great black piano appassionata' show that he is resisting and holding the flood of nostalgia, looking at it—this reader takes the poet to be indulging in it. In this he shows his own fear of beginning the process of recognising the nostalgic elements in oneself, and so, in his angry reaction, betrays an uncertainty of himself. His angry rejection prevents him taking in the actual meaning. The meaning of the poem would have been too painful, so he attacks it, distorts it by amendment, and so makes it fit only for comfortable rejection. He rejects a part of himself, really, which he has projected into the poem. It may take several reactive readings of this kind, before he is willing to allow the poem to 'work' on him—even to the extent of responding to it, in terms of even merely understanding it.

So, response to meaning in art, since art often deals with deep disturbances, often means overcoming resistances. We project over the poem some aspect of our inward life which is aroused, and fend it off by becoming angry, as did this student:

This poem is false. One worships the past in the present, for what it is, not for what it was. To ask for the renewal of the past is to ask for its destruction. The poet is asking for the destruction of what is most dear to him.

As Richards says, 'The writer is rather describing some other poem floating in his private limbo than attempting to discover what the poet is doing.'

Difficulties of responding to words—and of using words—then, are closely related to our whole make-up, and our ability to take in fresh experience, fresh insights and perceptions. With some things, we will take these in with great delight—gloating and exploring, finding great satisfaction, as when we watch a very beautiful sunset, or listen to the richest of Mozart's music. But there will inevitably be times, either in responding to art, or in reacting to experience, when we cannot accept without a struggle, because to accept the meaning is to require ourselves to make an adjustment in our personalities—and in our relationship with subjective and objective reality. Such adjustment is painful, and reminds us of the painful adjustments we had to make in infancy, to accept the truth of the world, as our phantasies proved inadequate means of dealing with it. We know that then those adjustments went with a fear of *not surviving* at times, and sometimes brought the bitter taste of loss and separation. We feel, as we mature, that we should conserve our attitudes, 'rest on our oars', and live by our comfortable, known patterns and acceptances of 'what reality is'. We do not like those moments, which come in sickness, over-tiredness, at big crises like birth or death, and in our deeper sexual life, which reveal to us that we have a continual 'reality problem', and must always be seeking to maintain our identity. We do not like the mind's 'cliffs of fall'—we like to be comfortably cruising along, with stabilisers and air-conditioning. Storms, changes of course, hard work to accept new perceptions and conditions, prayer and acute awareness—all these are taxing and harrowing, even though they bring great satisfactions in the end, when peace is achieved and landfall gained.

So, we avoid them where we can—and so we avoid art and creativity, where we can—seeking to turn art into dilettantism, or aestheticism, or 'scholarly' academicism, or exam stuff—into anything, so long as we can resist and escape the erosive and revolutionary flux of art. Yet we know, in doing so, that we are debarring ourselves from our greatest satisfactions, and sources of growth, towards being effective in the world—by 'coming to terms' with outward and inward realities as they are.*

This process is going on all the time in the classroom, because teaching is, inevitably, a creative process, unless it is turned into a destructive or dead mechanical one. It must be one or the other.

* This account of 'psychic creativity' is an abbreviation of the account of processes explored in the books listed on p. 275, and referred to in Appendix A.

Yet, though what I have said in the last few paragraphs is far from simple, the essential process, where English is concerned, *is* simple—it is a matter of responding to words: in the face of all the impulses and complexities I have been describing, to allow words to work between us and experience, to foster change and growth in our personalities, to enlarge our capacities to explore and take hold of reality, and deal with it effectively.

In literary criticism, of course, there are some superb examples of the exploration of meaning in relation to these processes. One thinks for example of the brilliant discussion by William Empson of a poem by Nashe. (I resist the temptation to quote at length. The book is readily available as a paperback.)* The effect of the best of such close criticism is to lead one back with renewed keenness to the words-on-the-page, whatever reservations one may have about some of Empson's ingenuity. It has not the wrong kind of irrelevance—only that of the inevitable uncertainties of dealing with products of the unconscious and the imagination. Much of Empson's work makes a useful antidote to school teaching about the nature of 'metaphor' and so forth. His own grammatical interest is complex and exact, and his syntactical ingenuity perhaps too far-fetched. But his is the best kind of linguistic interest, which worms out the richness of words, and opens out poetical associations and ranges which one has passed by without suspecting what lay beneath the surface. Once one's attention is drawn to them the poem makes a deeper and more disturbing effect—requiring painful apprehensions of the nature of transience and mortality.

How does this process work, as between teacher and children, in the classroom situation? Let us return to the lines from *The Ancient Mariner*. Here we need to set these lines in their larger context in the poem.

What is *The Ancient Mariner* about? And why (anyway) has the teacher chosen it to read with his children? Or, perhaps a better way to put the last question, why was he impelled, for reasons of which he was perhaps unaware, to choose this poem? We will accept—I think justifiably for, say, bright 12–14 year olds—that it is a very appropriate and relevant choice.

The teacher has chosen it, let us say, because it stirs him deeply and seems to him very beautiful. He knows it to be simple in diction, full of drama, full of clear deep feeling, full of 'things seen' which are very striking. And it has the advantage of a ballad form with short lines, a

* *Seven Types of Ambiguity* (pp. 25–7).

vigorous rhythm, and a narrative which makes rapid progress. He also knows that it exerts a magical fascination for both adult and child, though he may not consciously know why.

He will also choose the poem for many intuitive reasons—and here we come to an important consideration of the situation, and its implications for training. To know 'what will go' with children, and to find the right poem for the right moment, is a matter of what we call intuition. What I think we really mean by this is (as I have insisted) that most adults are naturally teachers, and naturally creative—and have the faculty to do the right thing for children, so long as nothing interferes with or blocks these natural gifts.

The Ancient Mariner, for related reasons, is an excellent poem for study, for students who are going to be teachers, as well as for children: in this adult and child again, truly, 'meet in the word'. Teachers need to know about the inward processes that go on between creatures, and while some explicit knowledge helps, deep imaginative experiences can help even more, for they convey a 'whole' knowledge to a whole person, and thus even to the less accessible areas of sensibility. We may learn a very great deal about inward processes, as they are in adult and child from *The Ancient Mariner*. This is so because the poem is about states of being. Let us consider the way a teacher handles this rendering of states of being, in the classroom situation, in which he has part of his own being—and the children have part of theirs.

Let us consider the adult's part in this 'meeting'. The teacher has his own complex make up. Much of this will be determined by heredity, and by his environment when he was an infant, in which the dynamic patterns of his psyche were set. His experience of his early environment will have determined the degree to which he can employ his creative faculties—such as being able to deal with children, with his own emotions, and his inward and outward reality. His relationship with children will depend upon how secure he is in himself, in his sense of his own identity, and his own maturity. I have known teachers so afraid of their weaknesses, by which they fear they might too easily fall into 'childishness' that they have maintained a stern and tense remoteness, making it impossible for their children to 'give'. Others, perhaps less stable, or less in control of themselves, have alternated disastrously between lapsing into childishness, 'getting on the side of the children', and then, when the children tested them to see if they were really adult or no, they would return with alarming and cruel suddenness into very severe adults—usually losing their tempers in the abrupt process, and assaulting children in an unfair and dangerous way. In doing so they often attack the more unbalanced children—those who most 'remind' them of themselves, and their worst weaknesses.

Only gradually does a teacher learn to establish a relationship between himself and the children, in which he may be found by the children to be an adult, when tested, and show himself an adult who can allow children to be children. Only in such a distanced relationship—in which the adult knows he is an adult, and the children know where they are—can love and sympathy be given, and creative 'giving' and 'receiving' take place.

The teacher, then, has his own problems—and the largely unconscious 'searchings' of children are bound to bring them to his notice. For they need to discover whether the teacher-adult can be trusted—trusted to be *there*, to give continuity—not necessarily a consistent continuity (inevitably, of course, children find that adults are not always consistent): but at best, one whom they trust to offer a benign tolerance of all their efforts, and not reject them. These efforts are towards the good, the positive, towards order, and fine awareness—and only where there is severe damage to the psyche are children in any way destructive or vicious. But all children are in the process of developing their reality sense, and grappling with their 'reality problem', and they swing between bad and good, constructive and destructive, learning to resolve these ambivalent impulses.

One meeting place between the imperfect struggling personality of the English teacher, and the incomplete and wrestling personality of the child, is in the word—in the creative work that is done in the impersonal context of the classroom. The creative contest with experience can excite and satisfy both. Of course, one can opt out of this essential relationship: textbooks, for instance, provide a substitute in meaningless time-consuming activity for this creative relationship.

A poem like *The Ancient Mariner* is at the other extreme from 'practical exercises': in exploring inward perplexities as it does, by establishing a felt world of symbolic experience, it changes those who read it. What we go through if we possess it is a revision of our deepest attitudes to other living creatures, and to life. The teacher reads it with the thirty living creatures before him, with whom he is in a very special creative relationship. They are impelled to 'give' to him, and to explore with him. In this sense the children *are* the water-snakes! And to them perhaps he is the Ancient Mariner—but he is also Another Creature, such as the Mariner destroys, and sees around him, at every stage. There is, thus, by the agency of the poem, room for a great deal of identifying —and, indeed, a class would not be experiencing the poem adequately, unless it felt, at the moment quoted above, the weight of a guilt sliding from every neck, as the poem is read and discussed. Yet the teacher, despite all the complexity of relationship, the depth of the contemplation of Being, all the identification and metaphorical activity of imagination and mind that is going on—the teacher need be doing no more than asking simple questions about the meaning of words.

In his loneliness and fixedness he yearneth towards the journeying Moon, and the stars that still sojourn, yet still move onward; and every where the blue sky belongs to them and is their appointed rest, and their native country and their own natural homes, which they enter unannounced, as lords that are certainly expected and yet there is a silent joy at their arrival.

The moving Moon went up the sky,
And no where did abide:
Softly she was going up,
And a star or two beside...

Her beams bemocked the sultry main,
Like April hoar-frost spread;
But where the ship's huge shadow lay,
The charmèd water burnt alway
A still and awful red.

By the light of the Moon he beholdeth God's creatures of the great calm.

Beyond the shadow of the ship,
I watched the water-snakes;
They moved in tracks of shining white,
And when they reared, the elfish light
Fell off in silent flakes.

Within the shadow of the ship
I watched their rich attire;
Blue, glossy green, and velvet black,
They coiled and swam; and every track
Was a flash of golden fire.

Their beauty and their happiness.

O happy living things! no tongue
Their beauty might declare;
A spring of love gushed from my heart,
And I blessed them unaware:
Sure my kind saint took pity on me,
And I blessed them unaware

He blesseth them in his heart.

The spell begins to break.

The self-same moment I could pray;
And from my neck so free
The Albatross fell off, and sank
Like lead into the sea.

Let us ask some questions about words, such as children might ask in class. Before they can be answered, however, the teacher must be able to read well enough. In his digs the night before, he must have enquired for himself what the poem is about (even though he is not going to make this explicit in a lesson). If he is lucky he will be able to get hold of John Livingstone Lowes's *The Road to Xanadu*: or perhaps he will read Marjorie Hourd's gloss on these very lines, which I have just rediscovered in *Coming into Their Own*, which explores this moment in the poem as a moment of the outburst of love and gratitude. But if he has been properly trained, he will go back to the poem to question the critical comment of others—and this will involve him in puzzling over the words which he least understands. In making notes on these he will be on the one hand preparing himself to answer pupils' questions, and on the other to make up his own mind what the poem is about.

Here the questions in his mind might be:

What is the force of the word 'moving', because we do not see the moon moving, usually?

Of course the moon does not stop: why then does the poet say 'no where did abide'?

Why is the moon 'she': or, rather, why do I feel the moon is 'she', to whom he 'yearneth'? Why is the moon so powerfully associated with 'rest', and 'natural homes'?

That there is 'silent joy' among astral bodies is a pathetic fallacy: but yet the phrase is very moving—why?

What is the meaning of 'bemocked'? And what is the symbolism of the 'April' coolness of the moon's light on the 'sultry' and 'charmèd' water?

What is the powerful force I feel behind the humanisation of the scene—the way the stars and moon become creatures along with the water-snakes?

What is the point of the sudden 'spring of love'? Why does the spell break? What is it that slides from his neck?

Note the rhythms of the moment, 'O happy living things...': this catches the movement of a deep stir in the whole being. The utterance is spoken by 'no tongue', but by the whole creature *unaware*: there is no possibility of mere superficial assent, or insincerity. Why does the rhythm come to life so, here?

From this pondering of local phrases the teacher can go on to make his own possession of the poem (though continuing to keep in mind the fact that he need not inflict exegesis on the children—and need only ask them *some* of the questions above, about words, if he asks anything).

This moment in *The Ancient Mariner* re-creates in us deep feelings of reparation, and of that giving forth which we are enabled to afford, if we can complete certain processes of development and maturation in ourselves. The poem is the product of this discovery in Coleridge. To achieve this feeling of harmony with natural life, and this capacity to 'bless', we need to be able to accept the bad in ourselves, and to find goodness and creativity in ourselves and in 'the object' (that is, the 'other' towards whom our potentialities may go out in relationship). If we can, then we shall feel 'welcome', inwardly rich. We can see the world of outside reality as one from which we may draw even greater richness by taking into ourselves aspects of its beauty and vitality, despite its imperfection, ugliness, destructiveness and awfulness.

In *The Ancient Mariner* the essential theme is guilt for a 'thoughtless' attack on life (i.e. thoughtless in the sense of Lear's 'O, I have ta'en too little care of this!') and fear that such an attack (which is experienced by all children in their phantasy) will be followed by forms of magical retribution from the natural world. In *The Ancient Mariner* the forms the retribution take are like those of a child's vivid phantasy: the poem's naïvety goes with this element—which is like that of folksong. The meaning of the poem, though almost certainly never explicit to Coleridge, is to be found in its symbolising of deep and universal states of being. We would see them as having to do with reparation, and the 'stage of concern' (see p. 163, and the books listed on p. 275). Yet in order to possess these unconscious elements of the poem we need do no more than read the poetry well and let it work upon us. The deeper aspects need never become explicit. When teaching children we can bring them to a deep 'theologico-metaphysical' experience as Coleridge might have called it without any abstract discussion at all—simply by good reading.

In discussing 'method' the aim must never be lost sight of—which is to let the work of art of a great and subtle mind and sensibility work upon our whole being, enriching it, and conveying into our lives a deeper sense of beauty and meaning. This, of course, was Coleridge's and Wordsworth's purpose:

...my endeavours should be directed to persons and characters supernatural, or at least romantic; *yet so as to transfer from our inward nature human interest and a semblance of truth* sufficient to procure for these shadows of imagination that willing suspension of disbelief for the moment, which constitutes poetic faith. Mr Wordsworth, on the other hand, was to propose to himself as his

object, to give the charm of novelty to things of every day, and to excite a
feeling analogous to the supernatural, by awakening the mind's attention from
the lethargy of custom, and directing it to the liveliness and the wonders of
the world before us; an inexhaustible treasure, but for which, in consequence
of the film of familiarity and selfish solicitude we have eyes, yet see not, ears
that hear not, and hearts that neither feel nor understand.*

Coleridge's and Wordsworth's purpose was to help build bridges
between inner and outer reality, to enrich inward experience, to extend
the limits of perception, to awaken feeling, and to open eyes to beauty
and joy. This is our purpose in using their poetry.

The verses quoted above are perhaps the climax of the reparative and
creative urge in Coleridge, obtained through those metaphorical processes which require 'that willing suspension of disbelief' (or in other
terms a recognition of the existence of unconscious phantasy and the
reality of the inner life, in symbolic terms). It marks a stage in literary
history after which the darker areas of experience could no longer be
denied: Blake and Coleridge lead on to psychoanalytical exploration of
man's nature.

Thus it is appropriate to use psychoanalytical concepts to help us
possess the poetry: but the poetry must remain pre-eminent. And if
children (and students) possess the poetry, they are in possession of the
symbolism, its meaning and its affect, which is what matters.

Now to look closely at the poem, as a teacher should. Why does it
fascinate so? What is it about the language? One cannot separate the
verses of *The Ancient Mariner* from the deeply moving and significant
hanging footnotes. For these Coleridge uses an archaic and simple prose
obviously derived from the Authorised Version. From this we may
suspect that he is to a degree being nostalgic, and yearning for something
of the innocence of childhood devotion and legend, fairy tale, and folksong. We all commonly associate religious teaching and its emphasis on
'good' with our desire to emulate the 'good' mother, and the simple
rhymes and Bible stories learned at her lap. In this prose and the religious nature of the poem these nostalgic elements converge.

If we are to interpret the symbols in psychological terms it would
of course be too crude to say that the Moon 'is' the mother and the
Stars her children, the siblings around her. But something of the echoes
of earliest object-relationship is here—the Moon is 'she', and the stars
are in their '*natural homes*' which they '*enter unannounced as Lords that are*

---

* *Biographia Literaria*, chapter XIV on the genesis of *The Ancient Mariner* (Nonsuch ed. p. 237).

expected': here is surely a most beautiful metaphor of birth, mingled with feelings of joyful acceptance of the birth of other creatures (at an unconscious level the siblings who may emanate from the mother). The moon symbolises all those redeeming, lovely, be-gracing elements associated with 'she'—with the phantasy image of the 'good' woman which the infant derives from the mother, and which remain in our adult minds, as an embodiment of maternal care, and inspiration—in terms of the 'ideal object'. The moon is often a symbol of 'the object' with whom we yearn to relate.★ It is also a symbol of our own capacities, by introjection, to 'make' this beauty-and-grace-bestowing creature an aspect of ourselves, for she is the inspiration of all our reparative creativity. Thus, from the moon-mother, who bestows her transforming light of assuagement, love and beauty, the protagonist himself learns to live and bless. That is, our first object *is* the mother—and she teaches us to relate to others, and to the world: to all objects.

The Moon is 'moving'—'and no where did abide': she is universal, and is not claimed by any abode, at the expense of the universe (or of the monistic infant self who loves and claims her). She is inexhaustibly altruistic, always moving, journeying and bestowing—as the 'perfect' mother is, in the infant's idealising phantasy. The moon-mother is omnipotent: and her presence is always to be felt in the universe. To her we, like the Ancient Mariner, need to 'yearn' in our 'loneliness and fixedness'—for her influence can release the 'fixed' hardness of heart, the fixed curse, and draw generous feelings out of our monism, towards acceptance, as she sheds her light.

The stars (siblings, and those who have a claim to life as much as we) both 'sojourn' and 'move onward' with her. They are born, they are in our life; all move forward in time, and all, since all exist, must be accepted. In the short hanging footnote here we experience the stage at which a child comes to accept the reality of other members of his family and the existence of others in the 'not-me': 'every where the blue sky belongs to them and is their appointed rest, and their native country and their own natural home...'. Outer reality may be 'allowed' to exist in its own right: an important stage in our discovery of our identity.

This acceptance of the 'rivals' to the (moon-) mother's bounty and affection gradually extends to all natural life that shares existence on

★ See the moon as such a symbol in D. H. Lawrence's *Women in Love*, e.g. in the chapter 'Moony'. And in Gustav Mahler's *Das Lied von der Erde*.

earth with us—however low or bad: by this acceptance we also accept the mean and bad, the hate and the guilt, in ourselves. The effect of this acceptance is that we feel 'a silent joy at their arrival', rather than envy.

Hate and guilt which originate in the poem in the first thoughtless (concernless) attack, are externalised in the burning drought, the angry sky, the shadows and the sea, but now are 'bemocked' by the moon-mother's benign goodness.The sultry threats are challenged and dissipated by her light. Over the 'sultry main' has been cast the hate of the protagonist-child, by the controlling rage of omnipotent magic ('charmèd'). Over this the moon-mother's light of love is able to cast the white wild cold springtime aspect of hoar-frost. This is, I think, one of the loveliest verses in English poetry, and its poetic complexity is great. To the Ancient Mariner, exiled by the consequences of his hate, destructiveness and guilt, the light of the mother's benignity is like 'April' (of home, countryside and England) on the savage and enchanted tropical sea. With April hoar-frost come spring flowers, despite its touch of placatory coolness: the burning tropical sea is hallucinatory, deathly, 'still and awful'. The moon's beams evoke the ecstasy and pain of spring and growth; the red sultry main is awful with mirage, threat and menace. The ambiguity is complex: 'bemocking' is benign, while the 'charmèd water' 'burns' with malignancy. The intense ambiguity is relished in our senses, in the visual conflict between the white flakes of light that 'falls off', and the 'awful red': between the frost and the fire; the still sultriness and the leaping cool light. The intensity of sensuous contrast is evocative of the conflict of spirits between the mother, with her 'softly moving' power to chastise and subdue, and the enraged infantile anger that seeks for omnipotent and universal destruction— hate and guilt which yet inwardly the soul seeks to overcome. Out of the conflict comes the new richness—a perception of the 'rich attire' of living things as realities in their own right, in their coiling vitality, uttering their flashes of potential worth. They give off a golden fire of vital assertion, by contrast with the sultry sterility of the 'main'— their unpropitious environment, the indifferent outward universe— and by identifying, we, with the protagonist, feel we can be more alive and whole too.

The 'ship's huge shadow' maintains an area of savagery: here is the actual province of the consequences of hate and guilt. The water snakes, with their independent existence, move between the moonlight of the quiet mother, and return into the shadow of the ship. By their vitality of generosity, of vigour and beauty ('And when they reared, the elfish light Fell off. . .'), they make communication possible—and thus make reparation possible—between the positive and destructive areas of the soul. It is the beauty of this interchange, of apprehensions and feelings towards life, that generate a solution that breaks the spell. The solution is a richness that seems generated from the 'April' light of the moon: for the water-snakes move in the 'tracks of shining white', 'rear', and return to being a glad richness and vitality ('coiled and swam') to the area of guilty shadow. Psychologically, we can say the passage re-enacts the discovery of the reassurance of continuity, of 'going on living', by drawing on the mother's continuing love that enables us to build up rich resources of identity, by accepting reality inward and outward. The water-snakes are the creative processes of the living heart and mind, magical and beautiful, God's 'creatures of the great calm'—the forces of harmony, reparation and peace in us. 'Unaware' the conflict is solved within, as this moment is reached: the consequence is an out-flowing of love, and the falling away of guilt and its weight of de-structiveness, hate, and the escape from Death. The release is no false solution: there can be no lip-service, as the Mariner's lips are sealed by the agony brought by the curse of destruction. But the impulse springs from where the discovery of out-flowing love—the capacity to be generously glad of the vitality and beauty of others—is made, in the unconscious 'spring of the heart'.

But his 'Death-in-life' symbolises our need, throughout life, to go on working on our problems of hate and guilt, as the Mariner has to go on telling his story.

Such an account of the poem is inevitably complex. Yet we need not try to give children anything of the kind. They can intuitively grasp at once, hearing the poem well read, what it is about. This will be well known to teachers who have read children's own poetry with their attention alert, or who have watched their silent intuitive attention to such a poem, when read aloud to them.

Before I discuss the presentation of *The Ancient Mariner* in class, however, I would like to discuss how themes in children's own crea-tivity link their deeper interests with the theme of that poem.

12

RESTLESS GOSSAMERES

Let me now try to show how we can discover important clues to children's needs for certain kinds of literature by responding carefully to their expression. Just as the Mariner's very survival is threatened by an inward destructive badness, which is outwardly symbolised by the killing of the albatross and its corpse round his neck, so are children at times preoccupied with the threat of inward hate that sometimes seems likely to annihilate them. When a child flies into a rage he may feel that his own black aggressiveness is likely to tear him to pieces, and destroy him, from within. Against this he has to exert all his constructiveness, all his powers to discover a continuing, whole identity. In doing so he is trying to come to terms with what the Ancient Mariner seeks to accept. Of course, the creation of a poem is an important part of this constructiveness: so can be the reading of such a poem.

Here are some children's poems which spring from the same kind of unconscious predicament. The first is very much in the mood of *The Ancient Mariner* (though the teacher had not in fact used this poem with this class previously).* The evanescence here, as of the tormented identity that seems likely to dissolve at any moment ('the wind blew through him'), is of the same kind of symbolism as Coleridge's 'restless gossameres'—and the whole hallucinatory quality of *The Ancient Mariner*. This little boy would have responded to much else in Coleridge:

We all look up to the blue sky for comfort, but nothing appears there, nothing comforts, nothing answers us, and so we die...†

The poem utters the rhythms of a felt terror, such as *The Ancient Mariner records*:

Real pain can alone cure us of imaginary ills. We feel a thousand miseries till we are lucky enough to feel misery...‡

---

* The teacher was David Schonveld, a student from the Education Department at Cambridge.
† *Notebook*, p. 188 (Nonsuch ed.).   ‡ *Anima Poetae*, p. 156 (Nonsuch ed.).

The Mariner has to feel his pain (and Coleridge had to feel the pain to discover the spring of love and beauty). Real misery is the passage to the lift of the heart, 'O happy living things...'. A child can, at a crucial moment in his psychic development, suddenly express the pain of apprehending the terrors of annihilation—experience a 'useful depression'—and emerge:

Black Skeleton

1. The knife-like cloud called Black
   Skeleton came too far down one day,
   The birds soaring into Black Skeleton
   Never seemed to come back.

2. It seemed as if it would lightning and
   light all the world up.

3. And then to my surprise, I saw a
   Flicker of light come from
   Black-Skeleton
   It was like a sinking ship
   In a raging sea.

4. The wind was roaring, the cloud rumbled
   Along like the blast of a cannon.

5. The whole world shuttered that moment,
   The leaden cloud was every colour in the rainbow.

6. And Black Skeleton thinned white.
   The wind blew through him.
   And the wind tore Black Skeleton
   Apart.

In places this poem is conventional:

The wind was roaring, the cloud rumbled
Along like the blast of a cannon

but it would be a mistake, in a no-nonsense impulse, to 'correct' it, and suppose that 'of course' the boy meant 'shattered' by 'shuttered', and elsewhere gets his effects by careless chance. He utters a truth he could not explicitly comprehend, and we must take his poem as a gift from his collaboration with his unknown self, such as Coleridge's was.

The test is to try such a poem on various groups of students: discussion tends to centre on certain significant phrases and on themes whose import cannot be denied. In the end, the more striking phrases are discovered to have a quality of imagery and rhythm that could only have come from the deep sincerity of engagement with an inward fear:

> The birds soaring into Black Skeleton
> Never seemed to come back...
>
> And Black Skeleton thinned white.
> The wind blew through him.
> And the wind tore Black Skeleton
> Apart.

The disturbing effects of these lines could not be faked—they depend upon evoking terms of annihilation, of loss of identity. (How, in this context, they recall 'and see those sails How thin they are and sere!')

'Black Skeleton' is the vengeful emanation that comes from the loved object who has at times been hated: he is the angry father's hate, perhaps, or hate of the father for which retribution is feared. He is destructive ('knife-like'), he is deathly and black, and he is hate which has 'gone too far' ('came too far down'). The child is writing in symbols of his own inward black hate which he fears may 'go too far' and bring revenge that will annihilate him. Of course there is also accurate observation of the outer world—as of the way birds disappear into clouds not to 'come back': but note that the phrase is 'Never *seemed* to come back'. The capacity of Black Skeleton to destroy by incorporation is unknown: this is doubly terrible (and brings to mind a child patient quoted by a psychoanalyst as saying she tended to have 'the dreadful dreads'). Even the vivid realisation of aspects of the outer world is impelled by the need to find objective correlatives for inward exploration.

The fear of annihilation is bravely encompassed and worked through, until the final crisis of the thundering (and beautiful) moment of obliteration. For at that moment, when the 'shutter' of oblivion should fall, there is a vision of radiant beauty that transforms:

> The leaden cloud was every colour in the rainbow.

Extraordinary that a little boy in 2B should have such vision! I used this poem for a film sequence in a television programme and the perspective of a savaged landscape was followed by a rapidly dissolving panorama of cumulo-nimbus thunderheads: the effect was as of an extraordinary dissipation of menace. And this is what the boy achieves: for it is now the threat of retribution for aggression itself that is 'laid' (like a ghost)—*by the poem*. 'Thinned white' enacts by its texture and thinness of sound and texture the desired evanescence; the placing of 'tore' enacts the constructive achievement of the poet. The aggression that threatened dissolution is embraced, and turned back on the fear, which is itself annihilated. It is a poem of great courage, in self-discovery, in 'coming to terms with aggression' and using it to hold off fears of annihilation from within.

Here is another prose poem by a child of thirteen on the same theme. Tom was an unco-operative little boy: but by hard work at free association, and by using various stimuli, a young woman student teacher drew this remarkable piece of writing out of him.*

Fire

Fire is not understanding; he is reckless and ruthless. He bites when you touch him, he is angry. Why? Who has upset him? Why does he roar when devouring one thing and purr when devouring another? He is a giver of heat but he doesn't want you to take it. The naked tongues of flame reach high into the sky as if searching for food.

He hates the wind and the rain, the wind makes him curl up and hide and the rain makes him spit in a fury of rain and smoke.

What makes him so reckless? Why does he find pleasure in destroying things? Why does he gnash his teeth in anger at metallic objects? He is so powerful, he stops at nothing!

What would we do without fire? He gives us our power, he cooks our food, he is our angry helper!

This piece of writing strikes one at once as 'Biblical'—because the symbolism is metaphysical in the way that symbolism tends to be in (say) Ecclesiastes, or the Book of Job, and because of the antithetical rhythm. It is interesting in discussion to see how long students who deny that this passage is about 'anything other than fire' can hold out, against those who are able to accept its metaphorical power, and the truth that it is about hate. In such a discussion, of course, the tutor's aim is to break down resistances, so that people may allow the poem to work upon them. But this can be done by letting students puzzle at one another and stumble upon the meaning themselves, in the dynamics of a proper seminar.

Those who deny the meaning of this prose poem are trying to resist a distorting statement about terrible aspects of the natural world. The boy sees that hate and destructiveness in human nature have their

* Miss Averne Shirley, a student teacher, at a progressive school.

correlatives in the savage indifference of fire: and about these he asks fundamental, but unanswerable, metaphysical questions. So, from this puzzlement comes its Jobian rhetoric.

A student who seeks to assert that the boy had simply seen a fire on the way to school and is writing about that can easily be confounded over the first phrase: 'Fire is not understanding...'. The word 'understanding' implicitly evokes comparison with the human intelligence, and so the fire is immediately personified: the next word is 'he'. Fire is a monster who is indifferent: has no capacity to understand by identifying and introspection, as we have. The child has sought to identify with fire, which can seem, as it moves, devours and roars, like a living creature. But, as he identifies, he finds no understanding there: the most terrible thing about fire is that it cannot identify with you, as you can identify with it. It is utterly without those capacities of understanding by which human beings are capable of compassion, sympathy, respect, kindness. How terrible it would be to be incapable of ruth! 'He is reckless and ruthless...'—why of all aspects of human nature does the boy pick on these, as attributes of indifferent fire? Students often find difficulty in attributing to a small boy the capacity to comprehend ruthlessness.

Here one has to make a psychological point: that we have all known ruthlessness, and coming to terms with this is one of our major problems. The discovery of ruth ('the stage of concern') is a necessary stage in capacities to live with ourselves and others (it is ruth that the Ancient Mariner discovers as the means to discover the richness of life in its reality). This discovery is an aspect of the 'depressive position', and the emergence of the rudimentary discovery of the difference between the 'me' and the 'not me'. At first, when it is angry or frustrated, the baby is capable of phantasies of total ruthlessness. Because of confusion between phantasy and reality, and because of its uncertainty as to what is 'itself' and what is 'other than itself', these phantasies (which are directed at consuming the mother) seem to threaten annihilation. But, of course, the more the mother is 'discovered' to be another, the more *concern is felt for her* (and fear is felt about the consequences of attacking her, and of her possible retribution). It was D. W. Winnicott who called this stage in the growth of consciousness 'the stage of concern', asserting that it is a positive stage, because it is the stage at which the 'object' and the self are discovered as separate entities. Thus, it is the beginning of our discovering the truth of ourselves in a real world.

By the mother continuing to reassure the child that he is loved for his own sake she enables him to allay within himself fears of her annihilation and retribution: she helps him to accept his own guilt and to discover the real world. Also, by her capacity to receive his love (really no more than his smiles and looks, his grunts, his simple bodily givings and gifts of contact) she enables him to make reparation, for the emptiness, the damage, he feels his anger, his aggressiveness, his *ruthlessness* may have caused to her. So every baby lives through the agonies of the Ancient Mariner, and continues to act out the reparative impulses of 'concern' ever after.

So, our aggression and hate are always the subject of fear, because when they were at their most ruthless they seemed to threaten to annihilate the object of our relationship, and possibly ourselves. Reparation (creative effort) is a continual attempt to resolve the threats of hate, and to overcome them by love, in order to resolve and strengthen the identity. Children, no less than adult creative artists, continually explore (as indeed we all do) the universal problem of whether they are 'good enough', have 'ruth' enough, to survive. The writer of this prose poem about fire is terrified because fire is utterly ruthless—as he was once. The fire about which he writes with such an excited rhythm is a metaphor of the hate and aggression within himself. The beautiful antithetical movement of the prose itself comes from his awareness of the 'contrary states of the human soul' of love and hate: out of the conflict he discovers the necessity to embrace the fire-like part of one's nature, and 'come to terms with one's own aggression': to resolve the ambivalence. Then, aggression can become one's 'angry helper', a rich source of assertiveness in the personality.

The infant's impulse to devour is the basis of the sadism inherent in all love. So the roar of destructive anger and the purr of the love of destruction are indivisible aspects of our inward life: Tom sees them both in fire.

Why does he roar when devouring one thing and purr when devouring another?

If the infant offers love, and it is not received, he may conclude that love —and giving—are dangerous and bad. So, giving has its own dangers: this problem Tom sees in fire too:

He is a giver of heat but he doesn't want you to take it...

Yet the need for contact, for the nourishment that love can bring, is always there: the tongues of fire are a symbol of the tongues of human desire, naked and hungry:

The naked tongues of flame reach high into the sky as if searching for food...*

The conflict between love and hate in the child is enacted by the dancing of fire under the rain: the rhythm of childish anger is perfectly caught by the next breathless sentence: the fire is a writhing, curling up, spitting child:

He hates the wind and the rain, the wind makes him curl up and hide and the rain makes him spit in a fury of rain and smoke...

Then, more calmly, Tom asks those questions about human nature which are more interesting than any answers will ever be:

Why does he find pleasure in destroying things?

'Why should I not...kill the thing I love?' Yet there are realities that the most consuming anger cannot obliterate: even fire can only 'gnash his teeth' at 'metallic objects'.

But even if it were possible to extinguish all anger, it would not be desirable: having looked at the recklessness within himself, and its 'objective correlative' in the fury of fire in the outward world, Tom accepts it:

What would we do without fire? He gives us our power...

The final phrase—'angry helper'—expresses with neat poetic economy the nature of a profound inward truth. The passage has all the depth of such metaphorical expression as the Book of Job ('Hast thou given the horse strength? Hast thou clothed his neck with thunder?), or Ecclesiastes. It is a very *English* piece of prose, belonging to the traditional mode of popular proverbial speech: yet it was uttered by a rather sad little boy at a 'progressive' school, for a student teacher.

A simpler, but related poem, was written for another student teacher by a slightly spastic boy. To such a child, possibly, the incapacity might seem to be a consequence of his own aggressive phantasies—a retribution for having hated, ruthlessly. So, to have to suffer a permanent incapacity for having been once totally possessed by rage, seems worse than death.

---

* Here there is the same erotic symbolism with Biblical echoes as in T. F. Powys's Mr *Weston's Good Wine*: 'Tamar turned, and sat looking in the fire. The flaming tree had died down. Tamar stretched both her hands over the fire. She wished to see, rising out of the hot coals again, that tree of flame. But the fire was sluggish, and refused to rise for her pleasure' (p. 81).

I fear not death alone
But fire
Fire burns ragingly
My nerves are uncontrollable
When I think of it.

His 'uncontrollable' limbs are filled with dismay, when he contemplates the raging fire of his own anger: the predicament is poignant. He may feel unconsciously that his spastic condition is a retribution for his 'badness'. But, of course, to write such a poem, is to come to recognise the dreadful fear obliquely, and so to help reassure oneself of being good enough to survive, despite that physical weakness which seems to be an outward and visible sign that one is not 'good enough to go on'.

A great many children's poems (and stories) are thus related to the theme of a poem like *The Ancient Mariner*—the theme of seeking by the discovery of beauty, of balance, stability, tolerance and order, to overcome fears of annihilation and inanition. Inward badness and guilt seem to threaten very survival: the poetry can be a means of discovering —'unawares'—the means for the soul to survive. The poetry is an attempt to resolve the identity.

The same fear of inward destructiveness continues to haunt the adolescent, at a time when the turmoil of feelings becomes, once more, difficult to live with. In Mr Jack Beckett's anthology of adolescent poetry, *The Naked Edge*, a girl of 14 writes:

Your whole mind filled to the brim
With hate, jealousy and anger.
It grows and grows until you swell
And overflows with the devil.
It burns an amber hole, black and poisoned
Through your soul,
Until anger, hate and jealousy win
By devouring your whole body up.
At last it rushes out full of poison
And deadly words.

The fear of a 'hole', and of hate 'devouring your whole body up', are a poetical expression of the adolescent's preoccupation with fears of inward 'badness' doing damage to the self and others—in terms which evoke infantile phantasies of the 'stage of concern' in early childhood.

The more intelligent adolescent still expresses such fears, even in a rather histrionic poem, even despite obvious literary borrowings.

Snakes and Ladders

My life is a game of snakes and ladders,
There is no set pattern; every move I make
Has a result, a beginning, an end,
And is scrutinized by those set ready to pounce,
To condemn, to ridicule, to tear to pieces.
Every day is spent rushing up ladders of nonentity
And crashing down snakes of reality,
Like black stone power-stations,
Or a forgotten thanatoid building of desolation,
Blocking, spoiling, deliberately concealing any pleasing view
Behind them, the view of which I crave for,
Like an addict craves his drugs.
Sometimes I come to a ladder:
A visit to russet Hyde Park in October,
Praise from my teacher,
A new Presley record,
An excursion to the theatre—
Selfish things, but what a joy they bring to me
In those precious moments before I am hurtled back to reality.

Dear God, will I ever be allowed to pass into womanhood,
Into an existence of responsibility,
Of happy living, of a husband and children?
Or will I be struck down,
Blinded by that unquenchable terrifying threat
That hangs over the world like a heavy black blanket of destruction.
Are you going to allow me to know, to feel,
To experience cold, hard-living death by radioactivity?
Am I ever going to bring up children
Without the fear of mocking them?
To bring them into this red-hot world
Sitting on an active volcano?
Inside me there are endless voices
Cawing in triumph like crows, black and ugly,
Crying 'Destruction!', 'Death!', 'Uselessness!'.
And asking one question—the vital question:
  Will I ever be allowed to grow up?

This poem by a grammar school girl, Janet, is uneven. Some of it is
rather conventional, and much of it is borrowed. She is exercising her
fascination with long words—'thanatoid', 'nonentity'. Yet it seems to

me a valuable poem for a sixth former to have written. It has its own elements of sincerity—the willingness to expose the childishness that finds satisfaction in simple things

> A visit to russet Hyde Park in October,
> Praise from my teacher...

And at times the rhythm is so obviously sincere and 'meant' that one cannot but be moved.

The problem explored by the poem is that crux of adolescence, trying to find proportion in one's experiences. Just as the adolescent is puzzled by his (or her) rapidly changing bodily shape, and learning to live with all manner of new physical powers in all their raw gawkiness —so, too, new realms of emotional experience distort the accustomed patterns of childhood. New desires loom—'to pass into womanhood': extended awarenesses dwarf the individual life—'a heavy black blanket of destruction'. Yet the sensibility remains childlike and with a childish sensitivity.

> Selfish things, but what a joy they bring to me...

'Selfish things' suggests that the inward child-self nurses the childish satisfactions, but somehow feels guilt about the joy in them. The adult world is full of tests, examinations, critical attitudes to the adolescent, who is still little more than capable of small tender childish joys within: yet the anti-creativity in the adult world makes the 'little feelings' seem insignificant: 'scrutinized by those set ready to pounce'. The new emerging identity, despite its challenging energy, has no more than small childish *dependent* resources ('praise from my teacher...a new Presley record...'—the Presley pleasure is that of dependence on a group cult) with which to resist the impingement of seemingly irrelevant and negative routines of school and work. The adults seem too anxious to condemn, to ridicule, to tear to pieces. Their reality seems to consist of 'ladders of nonentity': it offers no help to the need for 'entity'—for a whole new personality, in the face of the testing, anticreative environment. The emerging identity is being built round those 'precious moments', cherished timidly against a world which seems intent on 'blocking, spoiling, deliberately concealing'.

The poem is, however, full of the adolescent's challenging impulse, to break through, to a future of richness of identity and fulfilment in relationship. The voice changes in the second stanza, and timidity is renounced, in favour of a claim for 'responsible' discovery of potentialities, to live from a true self, without mockery. Thus 'Dear God' is not merely histrionic, nor self-pitying, nor a religious utterance: it means, 'when I look at the world you surround me with, I'm driven to the extreme of scorn'. 'Dear God' is a histrionic cry directed scornfully at the generation she is thrusting away.

> Are you going to allow me to know...cold...death?

Is this adolescent girl too histrionic about the bomb? Surely those of us who have known war would hesitate to say so.

For us all there is an inward price to pay for the Bomb, which this poem registers: it makes it all the harder to find meaning in one's life. The writer's problem is to begin to establish, towards 'womanhood', a developing personality, out of a childish sensibility, and an activity in the world of affairs which seems largely negative, and futile, seeming to her to foster 'nonentity' rather than 'identity'. Within her, there are reflections of the destructiveness of the outer world—reflections of the 'thanatoid', 'spoiling' aspects of her culture. And, more deeply within,

> Endless voices
> Cawing in triumph like crows, black and ugly,
> Crying 'Destruction!', 'Death!', 'Uselessness!'.

These are the common doubts of adolescence, with its tendencies towards schizoid episodes, when hate, futility and despair seem likely to triumph. Janet is writing to seek to gain hold over these dissociating elements in her personality: her poem, whatever its faults, is an attempt to express and control them.

But the bomb is an 'objective correlative' of the inward destructiveness. It is undeniably there, and actually poisoning us all with the evil consequences of man's capacities for hate. So, under the 'unquenchable terrifying threat', the outside complement to the inward blackness, the adolescent finds it that much harder to solve the inward problem, of coming to terms with the destructiveness within. As we have seen, this inward aggression can be made 'an angry helper': but the outward existence of the Bomb makes it seem almost impossible to embrace our own evil, so fissionable, so utterly rending and 'thanatoid' does it seem.

Of course it is typical of an adolescent, to bite off more than she can chew. The poem is mixed, and the inner and outer issues are such that nothing short of a poem of the stature of *Four Quartets* could begin to resolve them. But her poem is sincere, energetic and moving and it helps to make my point, that children's poems are often concerned with keeping the identity strong and growing, by overcoming inward threats of hate and guilt—crying 'death', 'destruction' and 'uselessness'—which is essentially what *The Ancient Mariner* is about.

13

FACILE CREDO...

So, we come back to *The Ancient Mariner* in the classroom. We have discussed the teacher's inward dynamics, and the poem's relation to these. We have done the same for its effect on the children's. We have tried to say what we think the poem may have done for Coleridge. We have analysed its symbolism of universal subjective preoccupations, and the way in which it resolves these with such beauty, so that possession of its phantasy helps release fresh potentialities in the self—yielding deeper insights into the heart of things.

It all sounds overwhelming: dare we ever teach the poem now? The trouble with exegesis of any kind is that it tends to raise such doubts. What we have to do is but to sing the song again: we must go back to the marvellous creation the poem is, and the marvellous thing our response to it is, each time we read it.

So, we do this in class. We shall ignore Coleridge's Latin foreword ('Facile credo, plures esse Naturas invisibiles quam visibiles in rerum universitate...'): and we plunge straight into a good story. We need do no more, on Monday morning, than open the book and begin: 'How a Ship having passed the line...'.

The present situation is a wedding: jovial and full of the bloom of youth: the bride 'paces' in:

> Red as a rose is she;
> Nodding their heads before her goes
> The merry minstrelsy.

But even in the midst of normal conviviality we are in death: the Mariner is under a compulsion to speak of the other world, while the Wedding guest 'cannot choose but hear' and 'listens like a three years' child'. So will our pupils, for the Storm blast is at once personified

> and he
> Was tyrannous and strong...

The world of the convivial throng, of wedding party or school, changes rapidly for another:

And now there came both mist and snow,
And it grew wondrous cold:
And ice, mast-high, came floating high
As green as emerald...

All we need, to begin with, is a good reading aloud: the intensity of Coleridges' realisation rapidly evokes in our senses the tactile experience of changes of environment. And with this, a sense of shift, into another sphere. And so, we move into symbolism. The ice *growls* and the fog-smoke is *white*.* The albatross seems a 'Christian soul', where it circled the ice splits 'with a thunderfit'. The bird perches for 'vespers nine' on 'mast or shroud'. From reminiscence of this other world the Ancient Mariner is still 'plagued':

'God save thee, ancient Mariner!
From the fiends, that plague thee thus!—
Why look'st thou so?'—'With my cross-bow
I shot the albatross.'

There are 82 lines in the first part of *The Ancient Mariner*: like the first act of *Hamlet*, it takes one immediately into the symbolic action, and the 'willing suspension of disbelief'. We are in another world from the world of 'merry din': we are made, compulsively, to recall a world of ominousness, of 'other' experience, primal and elemental: 'The ice was all between...'

From somewhere in this primal region we feel the burden of guilt: what was received 'with great joy and hospitality', what helped to split the ice, what perched 'for vespers', and was called a 'Christian soul', has been destroyed: 'The ancient Mariner inhospitably killeth the pious bird of good omen'.

In an attentive class there will be no need to dwell on the situation, the predicament: since everyone is in the predicament of feeling nameless guilt for primal 'inhospitality' (or ruthlessness). There could be that kind of rapt silence which indicates that children are already wholly engaged.

The poem certainly must be kept at its own level of evocatively tangible experience and phantasy symbolism. To ask 'What does the albatross stand for?' would be destructive of the metaphorical power

---

* There is a play, possibly, on *alba* in the word albatross, meaning white: there is much reference to whiteness in the poem. Whiteness is purity. The point symbolically is that the albatross was innocent, pure, offered love, attacked by hate for no cause. So whiteness must be refound, and is found in the white hoar frost of the moon's light.

itself. But what can be asked (though perhaps not just now, but at a second reading) are such questions as: 'What is a *dismal sheen*?', 'What does "The ice was all between" mean?' Some phrases, however, simply need sensitive tasting: they're plain enough to any Englishman, given their chance:

> And ice, mast-high...
> It cracked and growled...
>
> Whiles all the night, though fog-smoke white
> Glimmered the white moonshine...

Ice which *growls*, and is *mast-high*, is ice out of our experience: but conveyed to us in such simple language makes us aware of possible experience (certainly, we can come across 'muttering crags' such as Wordsworth heard, and growling can be heard in rocky mountain torrents).

The *fog-smoke* is one of those intense images whose sensuous force tends to make the bristles rise on one's face as one reads Coleridge. The consonants enact the choking nature of the fog, and the obfuscation of the senses. A *white* moonshine glimmers through *white fog-smoke*: there is blankness of the senses, as though the soul was choked, blinded and empty. The *smoke* suggests a fire somewhere: an anger, a threat of destruction, lurking in the 'land of ice, and of fearful sounds where no living thing was to be seen'.

It entirely depends upon the particular human situation whether or not the teacher dwells on the ambiguities of such words, and ponders their aura aloud. At the back of his mind will be the recognition that his pupils could tell him associations and meanings he does not know—and so illuminate Coleridge for him. Knowing as he does that Coleridge's poem is about primal guilt, he will find himself deeply moved by pupils' questions and comments—and by their rapt attention, when he gets it.

Yet he never need do more than keep the attention of English people on the meaning of English words. So we are back to our original problem:

> All in a hot and copper sky
> The bloody Sun, at noon,
> Right up above the mast did stand
> No bigger than the Moon...

'The sky's o'ercast with blood...' And yet it is not: something even more terrible is imagined: the tropical sun is small, but not white like the moon (that glimmers white through the white fog). It is a baleful red. And the sky is like a huge metallic vessel, hot, as if we were in the inside of a furnace. The world is sterile, hostile, uncanny: we await the witch's oils burning in the water, and the approaching barque of death. We are captivated, in hope of attaining release from our burden of guilt: of experiencing this for a moment at least, in phantasy. With such a poem teacher and child can experience together a profound spiritual experience.

Yet the ring is held all the time by attention to simple words: why *copper*?—just as in an art gallery we know we are looking at paint on canvas. So we go on, puzzling at the words, and making them yield up their full flavour, between poet, teacher and child, in the exploration of subjective experience, by creative symbolism.

Since I have dealt with practical methods of teaching poetry in the classroom so often elsewhere I need not go into details of method here: what I hope I have conveyed is the nature of the approach to 'meeting in the word' that seems to me our essential discipline.

14

NOT MEETING

How far short of being adequately trained for such an approach students are in colleges of education may be demonstrated yet once more in a relevant way. I have suggested that the adolescent Janet above is doing something very much like what Coleridge was doing in *The Ancient Mariner*. I set her poem in an examination, to test the capacities of a group of second-year men students in a college of education to 'receive' such a typical, genuine, adolescent poem. I also asked them to suggest works to which they would send the girl writer, to help her find sustenance, from literature, for her inward struggle.

The answers reveal a failure of their typical Eng. Lit. course (which on the face of it was a good one) to give students the capacity to 'take' imaginative writing and to make a positive response to young people's expression. Their course had given them little sense of how, having roused an appetite for words, and for poetic exploration of experience, they could lead such a girl to the resources of literature. Their answers consisted, as so often, of 'examination shadow-boxing', on the A level pattern, round the subject:

The first thing to ask ourselves in making any critical appraisal of a piece of writing is, 'what are the intentions of the author?' The next thing is: 'What degree of success has he or she achieved in conveying an idea or emotion?'

This student goes on to discuss with unyielding detachment questions of 'unintended ambiguity', and the need to

put ourselves in the mind of the poet, even to the extent of adopting his private language, and thereby to receive as nearly as possible the abstract idea or emotion he wishes to convey...

Why not, one protests, simply read the poem, respond—and say what you think and feel about it? Why shadow-box with such gloved detachment, as if to protect yourself from what seems a perfectly straightforward expression of a dilemma, of someone your own age? But no: the man must go on, about irrelevancies—the 'intellectual message':

12-2

The word 'reality' is one which has given to philosophers and philologists as much difficulty as any, there being no less than four quite distinct meanings in the Oxford English Dictionary...

'Stop showing off', one wants to cry. But if the student were to stop making his defensive display—would he have anything to say? Can he read the poem? It seems not: whenever he touches on the meaning his comment is inept:

In the poet's world she cannot win because there is nothing good and valuable about her view of reality...

I feel that working out her metaphor, she ought to allow for the fact that in time a player can finish the game and achieve the goal...

However, he says, with patronising generosity,

Allowing for the inconsistencies and for the nihilist view which Miss Janet takes, I like this poem and find it skilfully done in parts. Of course a poem cannot be good in parts; it should be considered as an organic whole; but one must not judge the work of an adolescent as one would judge that of an accomplished mature poet...

Such a reply reveals a responsiveness dulled by habits of suppressing one's true responses. Whenever habits of impressing the examiner by display have to be dropped, the student becomes confused. The writer can give us no indication why, in terms of total effect, he finds the poem 'skilfully done in parts'. He is so much on the defensive, distinguishing between 'unintended ambiguity' in adolescent poetry and 'conscious ambiguity and paronomasia' in the 'adult and sophisticated poet', that he cannot take the simple, direct poem at all. He really sees the impression of her anguish to live by the adolescent girl as 'nihilism': so, as we shall see, do many of his colleagues, and so they recommend literature for her on this basis.

The candidates were so influenced by the fashionable cults of 'futility' of protest that they could not see that what she sought was to *conquer* destructiveness. Since they could not take this positive meaning, they found it hard to think of appropriate literature to recommend her.

Another student does rather better than the first, but though he 'takes' the poem, he cannot allow that to be so moved is legitimate:

In the description of her feelings Janet has succeeded in creating something aethetically pleasing and emotionally honest...

but

Janet's poem is technically very good, but...attempts to express emotion that the writer does not feel...

Another student does get closer to the meaning in the poem, and in local analysis detects some of the obvious faults:

It will hardly do to claim that a heavy black blanket which blinds and which is unquenchable is a collapsed image. It is not effective but merely flamboyant...

He has the generosity to say:

As a record of the 'highs' and 'lows' of the sensitive adolescent mind this is a fine poem in itself.

But he does not say in what its fineness lies, nor does he come any nearer to its meaning than this. Another candidate does see the inward struggle:

It is a plea for sanity and order in the troubled mind of a person too young to properly understand the enemy she so bitterly fights against, or the friend that lies within herself...

But what seems to be an insight is merely a rather 'clever' phrase, for the writer goes on to make his display of equipment to 'appreciate'—destructively:

The words 'scrutinized' and 'pounce' continue the 'snake' metaphor but that reptile neither condemns, ridicules, nor tears to pieces—it poisons. I don't think the writer meant us to believe she is poisoned by life...

The tendency is to display a capacity for irrelevant exegesis—to fill an exam paper. Each of these candidates would have articulated their response quite differently—more relevantly and more generously—in a seminar discussion. But the writer goes on now simply to attempt to entertain the examiner:

I suspect the writer of pouting. I suspect her of flouting and of shouting. I suspect she makes incursions to theatres. I suspect she suspects...

So, from what seems a sensitive insight, this writer descends to a facetious destructiveness that the poem certainly does not merit:

'Any number may play,' but will the bomb hit only her pale pink counter?

and supercilious—'realistic'—irony at the expense of a sincere poem:

Our lives are real. The diesel fumes and the telly, the pay-packet and the parking problem, the two miserable weeks at Bognor, and the 'payments' are all real. We chose them. We continue to choose them. They are life. You can clear off if you want to (ref. Philip Larkin) but you can't escape reality—only hope to understand it...

This young man shows his incapacity to receive the poem, to examine it and compare it with his experience, and then to discuss it as the thing it is. He substitutes some commonplaces of his own, and merely preaches to the writer:

I would suggest some more objective experience, some hard work perhaps, some physical sacrifice, some danger, something unselfish, some job that can't be completed in an hour, something real...

The girl, he urges, should read the Bishop of Woolwich, Sir Winston Churchill, and Adlard Cole's *Pilot's Guide to the English Channel*. While a young man retains such arrogance he is unlikely to make a creative teacher: but how could his failure to answer such a question be conveyed to him?

It was possible to find, in a sentence or two, in perhaps one in every five of the candidates—no more—some indication that they could read the poem, and take its meaning, before giving an opinion of it. An older man writes:

I find the second stanza of the poem intensely moving. The sentiments themselves are, of course, common enough, and are fairly adequately summarised by the question, 'What's the use?'...She is afraid that her own life will be snuffed out before she is able to attain the mature life for which she yearns; and she is afraid that, even if she survives, her children will be mocked by being brought into a world of fearful terror...and utter destruction... Janet has had something to say and has said if forcefully and sincerely...

But the effect of the question—as so often when one sets such a test of spontaneous reading and 'receiving' capacity—was to reveal that these students simply had not been taught to read well enough to be teachers.

The replies, however, to the second part of the question were devastating. Here we are concerned with the germ of English teaching—where the growing child's life-needs can be nourished by fresh substance from the body of literature. If we know children, and can accept and receive their work, we can see how to help them dig out riches from literature, and absorb elements from it into their own processes of growth. What Janet needs is material which will reveal to her that she is not living through adolescence in isolation: that her perplexities about her own development are universal. She would understand, for instance, the 'Nick' stories of Hemingway, or she could follow the conflict of love and hate in *Wuthering Heights*. There are many young women in Lawrence's short stories with whose torment she could identify herself, as she could with others in the early stories of Doris Lessing. In various moods, and for various reasons, a sensitive teacher would earmark for her, Blake, Edward Thomas (and Helen Thomas's *World Without End*), Robert Graves, Edgell Rickword, Wilfrid Owen and Isaac Rosenberg from the First World War; but also *Hamlet*, and even Ezra Pound's translation of Sophocles's *The Women of Trachis*. Arthur Waley's *170 poems from the Chinese* certainly, because of her need to see a perspective in her life—'what, then, counts?'. And D. H. Lawrence's poetry, and T. S. Eliot's *Four Quartets*. I would hope to get her to read Jane Austen's *Emma* and George Eliot's *Middlemarch* to put some solid ground under her feet, solid ground of sense of personal value that can overcome hate. For this reason, too, perhaps, Conrad and Arthur Koestler's *Darkness at Noon*. Here is a year's work, at least, all of it substantially relevant to her needs.

Some of the sudents made quite positive and interesting suggestions —though nearly always conditioned by a sense that Janet's poem was merely about 'despair':

I should recommend her to read Louis MacNeice's *Fire* and *Springboard* where destruction and despair are looked straight in the face, but are in some way transformed. The fire during London's blitz becomes a beautiful tigerish thing, and the suicide wanting to hurl himself from the board above the street dies not in utter self-loathing but as a sacrifice...

We can dismiss the depression of Mr Larkin and build up a real vital world of D. H. Lawrence, where growing up is sympathetically portrayed without ignoring its poignancy and sadness...

Hamlet...is as modern as can be as the subject of independent decision. Eric Fromm *The Art of Loving*. Margaret Mead's *Growing up in New Guinea*.

But the generally superficial concept of the relationship between literature and life, and its 'moral' impact, may be represented by an older student, who makes what seem now rather old-fashioned recommendations:

It would almost certainly help her to realise that the horizon is not uniformly black if she were to read Evelyn Waugh's war trilogy—*Officers and Gentlemen, Unconditional Surrender*. From these she may learn that a sense of humour is never amiss in forbidding circumstances. She might, with advantage, also read some of Waugh's earlier satirical works.

Norman Collins is another writer...for a fuller understanding of the teeming life of London. *Bond Street* and *Children of the Archbishop* spring to mind...

Monsarrat's *The Cruel Sea*...*The Tube That Lost its Head*...H. E. Bates, and, more particularly, the Larkins...the gay, abounding, vitality of the Larkins will make a new woman of her...

It would seem to me that to direct a young person driven by such sincerity to trivial and commonplace works would be to betray her keenness and sensitivity. Another student would plunge her through the whole syllabus:

the works of Milton, who had to suffer the humiliation of seeing Charles II and his dissolute courtiers in power

> ...*the sons of Belial*
> *Flown with insolence and wine*...

...such a work as *Paradise Lost* where Milton identifies himself with Satan, and by writing this epic, so helps him to solve his problems, would certainly deepen her interest in literature and help her to see her own problems in perspective...

Alexander Pope was upset by the dissolute life in Church and Court... *The Prelude*...would serve to enlighten Janet further, while the problems of Coleridge in *The Ancient Mariner* provide the struggle within himself of Wordsworth's partner in *The Lyrical Ballads*...rollicking novels by Fielding...

Another indicates an odd clutter of signposts:

I would send [her] to the great interpreters of human nature: to Shakespeare, Chaucer, Defoe, Austen, Fielding, G. Eliot, to writers with a love of their fellow men...

—as the internal examiner has written, 'A bit broad'!

But most of the students showed no capacity for relevant selection at all: they showed no sense of having literary works in their possession, at their finger-tips, available for exact creative teaching purposes. What they did was to repeat the names of some of the authors which they had been told to read, for 'general literary knowledge' for A level—and beyond that they do no more than repeat fashionable names:

Dylan Thomas's *Under Milk Wood* gives a rich view of adult life, motherhood and other attendant joys...

(Surely there could be nothing wider of the mark about Llareggub!)

Koestler's *The Act of Creation* because this kind of pre-digested synthesised knowledge is useful to adolescents and not too didactic...

Colin Wilson's *The Outsider* might lead to some further reading...

I should recommend to her the poetry of Robert Lowell, Philip Larkin, Christopher Middleton, Ted Hughes and she could find interest in Sylvia Plath's *Ariel*...

I would recommend her to Kingsley Amis for fiction—but no time to mention more...

I suppose the plays like *The Chains*, *Waiting for Godot*, and *End Game* would fit in nicely with the girl's unbounded pessimism and theory of aimlessness...I think Sartre has the answer in 'human life begins on the far side of suffering...'

T. S. Eliot and Ted Hughes might be of value and they might perhaps go back as far as Rupert Brooke...Neville Shute, Sir Arthur Grimble, Nicholas Monsarrat and Sir Winston Churchill.

Philip Larkin, who might give her a slightly different slant on life, David Holbrook, who also comments on life in such poems as 'Whitsun Weddings' ...[!]

As she seems interested in God, Graham Greene, Aldous Huxley, or Evelyn Waugh...

Clive Bell, *Civilisation* might stimulate her to find a deeper interest in the arts...

C. S. Lewis for his religious profundity...William Golding, with his view of the inherent nature of man and his condemnation of torment might provide a salutary warning for her. Graham Greene with his deep concern for morality and his deeper understanding of evil would provide a firm philosophy with which to attack life.

How many of these writers have substance to offer the deeper imaginative needs of an adolescent? To engage in individual demolitions, however, is not my point. It is rather, that, since these comments are from second-year college of education students, what disaster has so disintegrated our concepts of relevant English literature that these few scraps of book review page values are all that is left, for teachers to offer a perceptive girl demanding a creative lead in living? How can one of these students write:

> Literature as I know it stops after Thackeray?

(Or perhaps that explains it all?) Why do they want to confirm her despair? Is it that they wish to endorse the unconscious destructiveness of *Take a Girl Like You*? Or of Iris Murdoch? In some cases, I think, this was so.

Certainly if teachers and pupils are to meet in the word, at the depth I have suggested in dealing with *The Ancient Mariner*, and to more creative effect, a completely fresh approach to the English syllabus in teacher training seems to be called for. If we look at what form this syllabus might take we shall see that it will have less to gain from being dictated by university English Departments than from being influenced by sensitive perceptions of what goes on in the classroom. Nor are the essential disciplines of meaning easy to explain to members of committees from other subjects, who see 'intellectual rigour' as being a certain kind of closed meticulousness of scholarship. The essence of our concern with meaning is that it should be open and creative—intuitive, at home with the unconscious 'unknown self' and freely flowing.

The creativity that may be found in every child teaches one what these disciplines are: so colleges of education which keep close to the observation of children are on firm ground. Teacher training has everything to learn from the living and creating child. The danger is that it has been inhibited by uncreative theories derived from mechanical studies of partial aspects of children (as if they were dead subjects), and by an overweight in the syllabus of subjects which move towards arid 'scholarly' investigation—such as the history of education or 'educational philosophy'.

Here, I think, we must briefly offer a show of resistance to the impulse —now found in many colleges of education—to substitute for a concern for meaning in the fullest sense, the disciplines of the 'new linguistics'. The new linguistics is obviously exerting itself strongly within the National Association of Teachers of English. In many ways it is a valuable discipline: the danger is that it may too easily become yet another substitute for that kind of engagement with language which involves us in discrimination between modes of experience and of behaviour—which is concerned with meaning and 'felt life'.

There are arguments in favour of linguistics. Where a 'language qualification' is demanded for examinations it could be less of a blight than the old grammar grind. In itself it is an interesting subject. Its new realism could help a new discipline to develop, which would refreshingly challenge traditional prejudices about 'correctness', and snobbish attitudes to forms of expression.

But enthusiasts for a new discipline never seem to be satisfied with the limitations of their subject. In looking at the syllabus it is necessary to urge that linguistics does not foster literary judgement, and only goes a very little way towards fostering the capacity to respond to meaning, in the full poetic sense. Its disciplines are not those of whole response.

While it does represent an advance in techniques of approach to the structures of language over traditional grammar, the new approach to language (associated in England with Professor Randolph Quirk) can never be a substitute for the disciplines of approach to meaning in whole terms, such as I have demonstrated above. Thus, we must make sure that students have sufficient time to exercise the intuitive faculties of exploring experience through language, by writing and reading, before giving up time to the formidable new specialisms represented by a syllabus such as the following (suggested by Professor Quirk for a new English Language Paper at A level). Of course, some students would find such work in linguistics rewarding. But it must not be made a substitute for the training in whole response to language.

A study of present-day English in relation to:
A. Its variety according to use... [subheadings omitted].
B. Its conventional nature... [notes omitted].
C. Its grammatical resources
  (a) classes and systems (that is, having a paradigmatic or 'choice' relation):
    (i) 'open' classes (nouns, verbs, adjectives, adverbs) as in 'the man came/stared/walked...'
    (ii) systems (the 'closed' classes of prepositions auxiliaries, pronouns etc.) as in 'The man will/may/shall...come.'
  (b) structures...
  (c) the relations of a to b (subject complement, head etc.) with reference to such categories as number, concord, mood, subordination, linkage, etc.
D. Its lexical resources:
    (i) paradigmatic relations;
    (ii) grammatical class correspondences (as between *advice, advise* and *advisory*; *help* and *helpful*; *dilute* and *dilution*);
    (iii) sets (such as *father, sister, uncle...tree, shrub, bush...*);
    (iv) series (such as *fourteen, fifteen, sixteen...*or tap water, rain water, bath water...).
  ...syntagmatic relations: (compounds...).

Such proposed papers testing knowledge in such matters reveal a subject which may be seized upon as another substitute for the development and exercise of judgement in responsive reading and writing. 'English' considered entirely as such work could become lost in the mechanics of language, plotting language construction in the absence of a concern for meaning just as much as the old grammar textbook approach. Already in America enthusiasm for the new linguistics and transformational grammar is generating language curricula containing exercises such as the following:

Construct kernel sentences which correspond to the following strings of symbols. Notice that the verbs have been supplied for you:

1. NP+ pres+ Vmid+ NP

|

cost   (*Oregon Curriculum Study Centre*)

No doubt the new approach to language structure has its own value and excitement for language specialists. But the fallacy is still to suppose that such a knowledge of 'how language works' can be the basis of expression, or promote the capacity to explore experience, to think and feel, to be articulate and effective in words, or to discriminate in a relevant way between works of literature. These demand a deep fostering of the inward capacity to find order and meaning, which is achieved by responding to language in its fullest, most ambiguous, most metaphorical way: a matter of whole and deep literacy, and yet largely intuitive. Any intellectual approach to language structure has something of an element of that impulse to control the subjective by the objective, in that it inhibits, bypasses, and seeks to tidy up the rich whole flux of language, tailing off, as language does, into penumbras and confusions, shady areas which are, however, useful to deal with the darkest and most intractable areas of existence. To the poet the confusions and ambiguities are valuable, since they belong to the puzzlement of inward experience itself. There are dangers in seeking to apply logic to the illogicalities of the unknown self: the inner world has its own poetic logic, which cuts across all categories and schemes that appeal to the intellect and ratiocination, defies investigation and baffles schemes. The poetic approach to meaning in language to the 'unknown self' (as Empson's book registers) is by indicating the necessity for ambiguity as an essential element of meaning. There is no substitute for the development of the grasp of language through the exercise of the disciplines of the exploring word, exploring inward experience thus, through phantasy and metaphor.*

---

* The opposite approach to language is that represented by Wittgenstein: 'What can be said at all can be said clearly, and what we cannot talk about we must consign to silence'; 'This means that the philosophical problems should completely disappear.' But the poet accepts that life-problems cannot disappear: the logician and grammarian often seem to be seeking to erect complete systems, whose completeness is only achieved by leaving something out, and so denying what can not really be denied.

15

THE EXPERIENCE OF
COLLABORATION

I shall suggest that to match the process of 'meeting in the word' in the classroom requires, in teacher training, not lectures, but seminars in which tutors and students meet as equals in the face of experience. To begin so with discussions of the meaning of words, in seminar work on literature, should suggest a completely different approach to the whole syllabus—at once more real and relevant. And there need be less anxiety about 'covering the whole field' (from *Beowulf* to T. S. Eliot) —for the student should be left not with a rag-bag of scraps of Eng. Lit. but rather a dynamic. That is, the student should have the capacity to take possession of a work, in such a way as to be impelled by interest and excitement to follow it up, in its human implications. This helps solve the problem of 'background': many English lecturers are obsessed with the need to 'fill in the background'. Yet this can only be done once the essential works take a meaning.

As an illustration, let us take a passage by Wordsworth, and assume we have seminar space to discuss it closely, and see where it leads us. In one of the training college examinations discussed above I was irritated to find one of the best students (student 2, pp. 70–71) repeating without any qualification a critic's trite conclusion that the Romantic poets, and especially Wordsworth, turned to nature for 'consolation' from the harsh world of man. Wordsworth surely cannot be understood as a great poet unless one finds in his work the depressive perception of a harsh indifference, and at times a horror, in Nature, as in the lines on the Simplon Pass:

> The immeasurable height
> Of woods decaying, never to be decayed,
> The stationary blasts of waterfalls,
> And in the narrow rent, at every turn
> Winds thwarting winds bewildered and forlorn,
> The torrents shooting from the clear blue sky,
> The rocks that muttered close upon our ears,

Black drizzling crags that spake by the wayside
As if a voice were in them, the sick sight
And giddy prospects of the raving stream,
The unfettered clouds and region of the heavens,
Tumult and peace, the darkness and the light—
Were all like workings of one mind, the features
Of the same face, blossoms upon one tree,
Characters of the great Apocalypse,
The types and symbols of Eternity,
Of first, and last, and midst, and without end.

If Wordsworth's attitude to nature is 'religious', then it is religious in the sense of awareness of the terror in external reality and of the reality of man's mysterious place in it. To speak of 'consolation' is much below the mark: one might as well speak of 'consolation' in Lawrence's religious attitude to experience, when he speaks of 'letting the fires of God blow through one'. A more adequate comment on Wordsworth's attitude to Nature is that of F. R. Leavis in *Revaluation*:

His mode of preoccupation...was that of a mind intent always upon ultimate sanctions, and upon the living connexions between man and the extra-human universe; it was, that is, in the same sense as Lawrence's was, religious.

(p. 165)

But one cannot grasp what 'religious' means in such a discussion unless, by responding to the words of poetic works, one does experience the religious sense. So, in order to discuss this matter at all (and establish between us what terms like 'religious' mean) we must *read the work* and 'let the fires of God blow through one', too. We must experience Wordsworth's depression before we can experience his triumph over it. Leavis will encourage us to do this: a poor critic will encourage us to funk it. But all the lecturer has to do is to encourage students to read the lines of poetry, keep their noses to the words, and wait for them to ask questions on their meaning. He will wait for students to ask, honestly, for help with words and phrases which they cannot understand. And he will try to discover 'agreed' meanings without implying there is a 'right answer'—certainly without implying that he can answer everything.
What does the line mean

Of woods decaying, never to be decayed?

Does it suggest the eternal cosmic patterns of decline and renewal, and the contrast between each living thing's individual decline with the everlasting renewal of life in general?

What is the effect of the phrase 'narrow rent'? With 'at every turn' this perhaps suggests being trapped in the toils of an indifferent nature, as in a narrow cutting between rocks. Discuss the phrase 'winds thwarting winds'. Why are the winds 'bewildered and forlorn'? There is a tumult in the winds often in mountain passes, caused by the uprushing airs in conflict. Wordsworth uses this as a symbol of the brute, unalterable conflict in the natural world, which, though it seems like turmoil, is yet more terrible—and can almost invoke one's pity, for insensate things 'bewildered and forlorn'. Students could bring these points out themselves. Then a tutor can suggest that at a deeper level Wordsworth is struggling to attain an adequate degree of toleration of external reality, and sees much in external nature of the mother and father, as an infant sees them, at the unconscious level, both succouring and threatening. Such psychological aspects in Wordsworth have been argued by Empson, Bateson and Leavis. The aspects of such unconscious elements which concern us in the art have to do with the degree of realism Wordsworth is able to struggle towards and achieve in his depiction of man's place in the natural world. So we begin to discuss Wordsworth on Nature and Man, beginning with the words on the page.

So, from the words, we begin to discuss attitudes to Nature, and so philosophy, and attitudes to life in history. Similar wide-ranging topics inevitably arise from the discussion of such words here as 'shooting', 'muttered', 'spake...as if a voice were in them', 'the sick sight', 'raving', and 'unfettered'. In their meaning, texture and sound a kind of contemplation of Nature is enacted which is far from 'consolation': so, we can dispose of the poor critic's cant, by which he has tried to push Wordsworth back into the murk from which he struggled to emerge by winning deeper and more painful insights.

No one could study such lines as these with care and still be satisfied with the phrases repeated by students in their essays on Wordsworth, borrowed from inferior critics in 'looking out essay material'. They might, on the other hand, arrive at perceptions of their own—if they exercised their voices on such passages week after week (and as they will have to do in school when children ask for meanings to be explained).

In some examinations I have countered the tendency to unload phrases from literary histories, by giving for discussion lines which revealed Wordsworth's perception of a Nature which is destructive. Several students were sensitive enough to see this themselves and were able to articulate their account in their own voice.

For instance, in discussing some lines from *Margaret*, student 34, as we have seen, said (p. 121):

Instead of just drooping and straggling the plants are *beginning to destroy each other*:

> ...*twined about her two small rows of pease*
> *And dragged them to the earth*...

Such close attention to lines of poetry elicits from students themselves a much closer perception of what Wordsworth's feelings for nature are than anything they gain from lecture notes or second-rate books of literary comment, by critics who seek to thrust back Wordsworth's art into comfortable stereotype.

What student could repeat that trite nonsense about 'consolation' derived from Nature, when she reads in *Michael* the following lines?

> There is a comfort in the strength of love.
> 'Twill make a thing endurable, which else
> Would overset the brain, or break the heart...?

Nature in *Michael is* indifferent: one feels it in the line:

> And never lifted up a single stone...

—only human sympathy can make such tragic suffering bearable. We could take another passage from *The Prelude* whose imagery is that of a brutal senseless nature that threatens, and only 'ill sustains' human life:

> Oh when I have hung
> Above the raven's nest, by knots of grass
> And half-inch fissures in the slippery rock
> But ill-sustained and almost (so it seemed)
> Suspended by the blast...
> Shouldering the naked crag...
> With what strange utterance did the loud dry wind
> Blow through my ear...

—'ill-sustained', 'almost (so it seemed) suspended by the *blast*...'. Attention to words here, their texture and meaning, would have produced a feeling for nature that would expose most of the generalisations in second-rate histories of Eng. Lit. ! Too often, however, students reveal in their work that they have not been taught sufficiently to look closely at words and so they cannot qualify their 'background' generalisations.

So, time needs to be given to pause, and to consider the more pregnant words in such a passage:

Ill-sustained. 'Sustain' (as in Shakespeare, 'our sustaining corn'—*Lear*) has the associations of cherish or nourish. While Wordsworth is here speaking of a wind holding him up from falling off a mountain, the word has implications about Nature's 'attitude' to 'sustaining' her creatures: can one feel a one-ness with indifferent forces?

Shouldering the naked crag. Here the image is not only of the bare rock, but also of the boy's body among its bareness. This conveys the physical feeling of being on a rocky outcrop: of being 'shoulder to shoulder' with nature, and a vitality shared with a natural world that offers nothing but insecurity (mere knots of grass) and a strange indifferent creativeness ('the loud dry wind Blow through my ear'). The passage enacts *combat* with Nature, the need to expose oneself to the reality of nature, and the fear inherent in this.

Once the content of such poetry has been grasped students can be allowed to generalise more widely under the control of their local engagement with the poetry: and here 'background' becomes relevant. *The Ancient Mariner* or *The Simplon Pass* can be discussed in their relation to the concern which produced *The Lyrical Ballads*. What is involved is a view of man—and this can be linked with the philosophy of the time (Rousseau) and events (the French Revolution). Once we have entered into the poet's feelings about childhood, man, and his inward life, biographical details become exciting—such as Wordsworth's relationship with Annette Vallon, and his sister.

Now, all I have said about 'meeting in the word' has implied a different approach from that from which most syllabuses are devised. That is, I do not begin from an assumption that there is a body of information—about literature, or human beings—in which 'right answers' are to be found, and all we have to do is to inject the information into students, who can then in turn inject it into their pupils. I begin from everyone's darkness.

All we can have, 'at the end of our exploring', is a dynamic. That is, all Coleridge achieved after creating his poem was a slightly deeper insight into man's inner world, and a sense of having for once grasped something of profound meaningfulness and beauty in life. All we can expect for ourselves is something similar, and much lesser: yet it will count for nothing unless it flows into dynamics already working in us. I have tried to show, in discussing *The Ancient Mariner*, how a poem reveals its quality by the degree to which it engages with the inner dynamics of the adult who is seeking by the same kind of symbolic 'work' to fortify his identity and to deepen his capacity to deal with the world. If the adult is a teacher, he will employ such works to help foster the inward dynamics, the searchings of his pupils, while they are in his creative care.

So now we are back with the small boy putting up his hand and asking, 'What does "bemocked" mean, sir?'. What happens at this point?

I do not think there is any satisfactory answer in theoretical terms: what happens will depend upon the actual situation between child and adult. Of course, as the author of books on teaching English I cannot pretend that one cannot be helped by reading about someone else's teaching experience. But I do not believe the teacher will be able to answer his pupil satisfactorily unless he has had a good deal of experience of actual give-and-take about meaning in his own education.

That is, though I should not be interpreted here as laying down any law of procedure, it might well be that the best answer would be: 'What do *you* think *bemocked* means?' That is, the answer to questions of meaning are most likely to come not from dictionaries or notes at the back, or even books of literary criticism—but from within the child and from within the teacher.* If we can respond to meaning (meaning goes with archetypal symbolism as in *The Ancient Mariner*) then it should be already possessed by us: what we can do is to deepen it by making it more articulate, by dwelling on it, and by exchanging possibilities and perceptions. How this can be done in practice I have discussed elsewhere, where I discuss the teaching of poetry.

We thus start, in classroom, or in seminar, from the collocation of responses—from the touch of the creative word on common experience. The astonishing poetic power of children's own expression shows that it should be perfectly possible to work in this way—*if* the teacher is properly trained in knowing what to draw out, from the flavours and effects of words. And the only way of training him is by giving him a

---

* See the children's answers to questions on the meaning of poetry in *English for Maturity*, pp. 82 ff. and those by less able pupils in *The Secret Places*, pp. 208-9.

rich experience of this activity—in teaching practice, in seminars on literature, in seminars on his teaching practice, on children, and on children's poetry.

My approach to the syllabus, then, will not be based on accounts of the 'perfect lesson' and the 'ultimate reading': but on an insistence that all learning in English must be based on collaboration, in meeting in the word, beginning more or less from scratch every time.

If we are prepared to work in such a way the development of our work will become a matter of seizing opportunities as they arise from experience of both life and literature, while hoping, of course, in general terms, to cover certain ground. If we work in such a dimension there seems little point in trying to set out a formal syllabus for the training of an English teacher. What will count is not the scheme of work but the experience of education that the student lives through—including, of course, his experience of the institution. So, rather than considering abstractions such as 'the history of English literature' and how to 'put this over to students' we need first to look at the living experience itself—*their* experience of *our* artefacts.

In this I have taken a first step, in looking at the content of students' essays, and of college of eduction examinations. These reveal that the cultural processes at some colleges of education are simply not good enough. They also reveal that in order to become a good teacher a student has to try to unlearn, and to live down, all the habits which O level and A level G.C.E. have conditioned in him. The experience of listening to 'information' lectures, covering a 'broad' syllabus, taking notes, writing these up into essays, memorising these essays, and scribbling them out under examination conditions not only gives students a merely superficial acquaintance with Eng. Lit.—it fosters dishonesty, as we have seen.

We need, then, to look at the lecture as a cultural feature itself, at the time-table based on lectures, at the kind of syllabus based on lectures, and at the essay, as artefacts which perhaps need reform.

I do not intend to propose a complete system: I shall simply suggest that reform requires beginning again from a completely different perspective—and in practical terms, from the seminar, in relation to students' experience of children. It was disappointing, to see in the recent White Paper on University Teaching Methods, that the seminar was confused with the tutorial ('seminar' taken as meaning 'a tutorial with multiple tutors') and given only one short paragraph, suggesting only that 'more use should be made of this method'. In teacher training, I am sure, the seminar is the crux—not least of the fresh and lively experience of an educational process itself—that student teachers need. What I mean by a seminar, as will be seen, is a discussion, led by a tutor, that begins with the students' exchange of their own experience, and their responses to a literary experience.

What we should start with is the hesitant, groping, at first inarticulate exchange between a small number of students, discussing the meaning of words with their tutor. From there, of course, the tutor knows where he is going—where he hopes to go, rather. He wants to bring his students up to be gazing over a wide prospect of the country of literature. But he wants them to be in possession of the characteristic entities that compose that topography (for its 'landscape' can only be an inward possession). So, there will never be a complete map—never a syllabus to be covered: tutor and students will be for ever mapping out and exploring what I have called indications. But the bearings, the experience, the sense of relationships and significances—if possessed—will be alive. In such a situation there is no reason why, from time to time, a tutor should not give a lecture, consolidating, and defining. But he will not go on, with any journey, until a work, or an author, is *known*.

There needs to be some kind of map of the ground ahead. But experience of teaching literature gradually reveals a tendency in oneself as tutor to prepare a syllabus which is ridiculously comprehensive—far beyond the capacities of any students. One tends to issue booklists to back these syllabuses up, as if with the commendable intention to 'stretch' students. Faced with a wide ranging syllabus and an overawing booklist, the student falls back on cribs, mugging generalisations, and short-circuit work.

Examined simply in terms of pages and hours to cover them, most syllabuses offer a number of texts which is unrealistic. Rough calculation reveals that if these syllabuses are to be covered a book such as *Emma* or a poet such as Wordsworth must be 'disposed of' in one or two fifty-minute periods. One cannot really 'teach' *Emma* or a proper appreciation of Wordsworth (say) in less than four two-hour periods, or, say, six fifty-minute ones at least. Even in six fifty-minute periods one could only *begin*—by, say, reading and discussing, say, the Lucy poems, 'Surpris'd by joy', *Margaret* and perhaps *Michael*, leaving no room for extracts from *The Prelude*, or to deal with some of the *Lyrical Ballads* most useful in school (e.g. *Goody Blake and Harry Gill*)—

or to discuss Wordsworth's affair with Annette Vallon, his feelings about the French Revolution, his attitude to Nature, his relationship with his sister Dorothy, or with Coleridge, or contemporary critics' opinions of him.

This over-optimism in 'covering ground' often goes with the belief that it is 'academic and scholarly' to be 'wide-ranging and extensive' in this way. The trouble is that, as we have seen from some of the examination answers, the 'range' lacks depth, and, too often, even sense, since the works have not been possessed.

The extensive 'outline' syllabus tends too to be accompanied by booklists recommending those books which aid the superficial possession of 'general' information about authors, and facile matter to use in exams, rather than books which oblige the reader to engage with the text. Thus, even in a women's college in which assessments have replaced examinations, a girl's bibliography at the end of her 'piece of work' on George Eliot contains no reference to *The Great Tradition*. Everywhere students are still sent to Elton, Compton Rickett, C. H. Herford, for 'literary history', and for particular writers to Hough, V. S. Pritchett, or Walter Allen, rather than to T. S. Eliot, D. H. Lawrence, L. C. Knights or F. R. Leavis. The reason is not unrelated to that of the impulse, discussed earlier, of avoiding the necessary pain, of response and growth, that attends creative reading.

Having indicated this central discipline of attention to meaning, there is no need for me to go so far as to draw up a complete syllabus for the training of an English teacher: all that needs to be worked out is the kind of organisation in which this true collaboration may go on. The central needs seem to me to be:

(1) Seminar work on poems, short stories, plays and novels, as the basis of all English. In practical terms this means groups of not more then twenty-four students meeting at least three periods a week of two hours, 'beginning from the page'.

This means a realistic re-examination of how many works can possibly be read and discussed. In my experience a substantial short poem such as Keats's *Ode to Autumn* or Wilfrid Owen's *Miners* can take a whole two-hour period at the beginning, while it should help foster the reading of a further six to ten poems in students' own time. A novel takes four to eight two-hour sessions, a play like *King Lear* at least five. For a year's syllabus to demand 'at least eight novels, eight plays and a comprehensive anthology of verse' is an unnecessary enumeration, hard to live up to in the event. Two novels and one play adequately studied would be enough. Similarly to take a (historical) 'two periods of literature' is too much for a first year's work if it is to have value in depth, and 'the works of Shakespeare' which some syllabuses announce should surely be amended to 'some works of Shakespeare'! *King Lear* or *Macbeth* alone really require many hours of application to understand: to believe anything else is to underate Shakespeare. As things stand some students probably have to 'do' Wordsworth in two or three fifty-minute periods. I would consider that the minimum time required to make the merest introduction to Wordsworth's *poems* is five or six fifty-minute periods, or, better still, three periods of two hours each, with plenty of reading in students' own time.

The work in English literature in some colleges has to be done in five fifty-minute periods a week. In one year for which I examined, three of these periods were devoted to 'the Romantic period' and two to 'the Victorian period'. This kind of division seems too mechanical (and tends to go with the stereotype answer). Far better to leave the intention to 'cover a period' as a general hope, while making the best possible selection of poems and other works to fit the hours available—which should primarily be devoted to 'reading together'.

What else do we need in the syllabus?

(2) Free drama, movement and creative writing: two periods of two hours a week.

(3) Experience with children—informal visits, work with youth clubs, periods of watching other teachers at work or helping with school projects (e.g. helping to produce school plays), etc.

(4) Experience of other creative work, in drama productions, music, painting, opera, sculpture, etc.

(5) Oral work and 'practical exercises'.

Nor must we forget (6), the study of the world of literature specially written for children.

If the creative disciplines are true disciplines then claims must be made for ample time and space for them, and if necessary these must be championed against the claims of psychology, educational theory and linguistics. Co-operation with these subjects, in ways indicated by this book, would be even more fruitful.

Thus, the main concern of the whole approach, starting from 'meeting the child in the word', depends upon there being:

(A) Much more teaching practice and experience of children.

(B) A close, friendly and *equal* relationship with the staff, as between adults and young adults, not as 'staff' and 'students'—that is, where work is concerned.

The first point is perhaps underlined by the following comment by a girl teacher about the lack of teaching practice at one of the supposedly more progressive colleges:

I think that perhaps the most important thing lacking in our college training was contact between students and schools. During our three years at college we had but two periods of teaching practice—four weeks in the second term of our first year when we were plunged in headlong into the classroom to flounder and either sink or swim—and eight weeks in our final year upon which we were assessed. This meant that apart from a rare visit to a nearby school our second year gave us little or no contact with schools at all.

This seems to me ludicrous. A nurse learns her job by doing her work actually on the wards—handling and nursing patients from the start. Though she is consequently given little time to study she is learning her job in the most valuable way—the method which we were given to believe, ironically, at college to be the most valuable of all in both teaching and learning: namely 'learning by experience'. Similarly in any other skilled jobs whether a profession or a trade the student learns by close contact with the job he is working towards—in any, that is, except teaching.

My own feeling is that a term's teaching practice each year is the minimum: and that six months would be even better. A related change —and no doubt radical change in some places—should be that the staff of a college of education must be expected to do regular teaching too. I know that one professor of education does this, and gives his staff opportunities to do the same. Such changes would ensure that teacher training keeps its feet on the classroom floor—most young teachers complain that their training was not realistic enough. It would also mean that the more intensive and deeper studies in English literature for its own sake—say of the poetry or novels of Lawrence, or George Eliot—would gain in maturity and depth. An additional advantage might be that colleges of education would attract fewer of those people who seek by such a post to escape the teaching situation in which they were not happy—and perhaps not very successful. If lecturers are meeting students regularly on terms of equality in seminar work on literature, and both are engaged in actual teaching, these experiences themselves would surely make it impossible for barriers to be maintained and themselves generate sympathy and understanding?

It seems strange, too, that, since they are concerned with the training of people who are to look after children, that training colleges should be generally so barren of children and their 'atmosphere'. Here, I am sure, there is much scope for experimental work. A most valuable development at one centre is their 'study practice' work on children: each student is given two children to get to know, to investigate, to meet their parents, study their work, teach, take on outings, and so forth.

Organisational changes to bring about such child-centred teacher training in colleges would, however, need more experienced interpretation in practical terms than anything I can offer, and I must confine myself to content, leaving the implications for organisation to be taken by others.

Changes in content and towards seminar work do, however, require some organisational changes, as suggested above (in B). An informal and equal relationship is required between staff and students for such an approach, and genuine seminar work would surely bring such a relaxed atmosphere? Some college of education staff still seem to assume an authoritarian superiority which divides them off. In one college one of the best students wrote:

I found opportunities for discussion on the whole more limited. There was an aura of importance surrounding the tutors that made one reluctant to venture any comment lest it should be thought ignorant. There were the exceptions, of course, but I missed the informal arguments we had had during lessons at school and afterwards, when we would linger behind to discuss a point of view with great vehemence... One still tends to feel one is being 'talked at' far too often. I should far rather read and make notes from a book, and then discuss and argue at a tutorial what I myself had analysed and digested, rather than blindly scribble one person's theories in a lecture with little or no time to discuss or dispute them...

Where this is a true picture of the situation, it is time things were changed: the best experience of education is the experience of a shared humility between teacher and taught in the face of life's perplexities.

Another major change required by a change of content is the abolition of examinations. Of course, it is possible to have tests which are genuine tests of useful disciplines. But what must surely be overthrown is the tyranny of examinations in departments and colleges of education because they are unnecessary.

On the question of the abolition of examinations, of course the reader should not assume that I am against the disciplined writing of thoughtful discussions of educational topics. My objection is that pieces of work for assessment train this faculty properly, while an examination manifests against such thought and discipline.

It is of the greatest value for students to commit their opinions of literature to paper, to read and consider authorities, to sift and judge the critical opinions of others, and to be faced with having to complete a substantial written study. But such work only has value if it increases the capacity of the student to read, to select from his reading, to judge it against his own experience, and to organise and write out his own compositions. Such work is only valuable if the student is encouraged to make it his own, finding his own way about the library, and making his own readings of verse and prose, albeit under guidance, employing critical terms in his own way, giving his own opinions, in his own words. Reading one's essay aloud to a tutor and discussing is an excellent method: possibly conditions hardly allow this system in most colleges.

Anyone who has compared the work of those colleges which take an examination, and those others functioning by assessment, as I have done, will have been struck by the difference throughout the whole approach to the subject.

I cannot avoid seeing the contrast between the systems of assessment at two centres where I was examiner as an object lesson. The 'pieces of work' (assessed) had a quality which I felt it was a pity the students who had an examination had not been able to experience. That is, I found that the 'assessed' students had tackled the writing of a thesis for their advanced main and main English studies with surprising skill. They had learnt to assemble material, to read and take notes, to ponder their subject, and organise a written study. They mostly wrote clearly and well, and went deeply into their subject, using their own perceptions and their own voice. It is no exaggeration to say that from some of these pieces of work a reader takes in fresh perceptions of an author (this was particularly true from the work of one mature student on Hopkins). As an author I could see how the preparation of such a substantial study had left its author forever changed and enriched by the experience. Several of these theses enriched one's sense of the value of literary studies—if only by the effort and sustained interest they conveyed.

By contrast the exam papers seemed on the whole irrelevant and dead —destined for the dustman in any case.

Except for a first novel by one of these students there was nothing at this college comparable in the work of 'examined' students, and in this they seemed by comparison deprived of something rich and valuable. They had had to write hastily in an examination, in snatches—unloading all they knew, with too little time to ponder and reflect. It seemed to me, reading their examination scripts, that it was wrong of an educational institution to demand that a student should write in this hasty and superficial way. Of course there were remarkable achievements (such as one man's answers on George Eliot and Blake). But the hasty handwriting, and the breathlessness of style, made one feel that the men had been submitted to a humiliation—to write a display, in facile terms, of their acquaintance with things about which they ought to feel too deeply to do so.

The abolition of examinations should also make more oral work possible. Those who come to the college of education as students often do not receive in their grammar school education sufficient training in literacy, fluency, and articulate expression to have a 'voice' of their own. They cannot express themselves clearly enough for the simplest classroom purposes. Their writing is stilted, clumsy, stiff, repetitive, illogical, cliché-ridden, callow and constipated. Though it is evident that the colleges already do a good deal of free oral work it is probable that—for the sake of *all* subjects, not only English—much more needs to be done.

Students should practise such oral exercises as I set out in *English for Maturity* (p. 129), using tape recorders and such apparatus. The basis of such work should be oral invention of stories and dialogue, the free writing of poetry and fiction, the invention of mimes, dance, free drama and scripted drama. In this the college of education obviously has to give a fundamental imaginative-creative training in literacy which the grammar schools have failed to give, because of the tyranny of examinations. Unless we make each next generation of teachers literate, this pattern will never be broken.

Creative work, oral work, and free association (in collaboration with the other arts) should provide a basis of fluency, as in the best English work in school. I assume these practical implications will be taken to follow from my emphasis on attention to poetic meaning. Students must be more trained to pause on words, ponder their emotive associations, their sounds, their complexity: and all manner of creative work aids this capacity, by freeing the imagination and extending perceptions of the inner world—and other possible experience.

There are other points which need to be mentioned, to make sure it is clear I do not mean anything vague or sloppy by creative disciplines. For instance, for a student to be able to discuss literature in the simplest terms in the classroom, he needs to have learnt some common critical terms—which he can learn best in seminar conditions, so that he can use them with confidence. Later he will have to do something even harder —which is to translate such terms into his own words. But he cannot do this without possessing them first. Here are some of the terms without which one can surely hardly begin to discuss literature:

Rhythm, metre (and the concomitant technical terms iambic, trochaic, anapaestic, dactylic, blank verse, heroic couplet, etc.),* movement, ambiguity, texture, enactment, assonance, alliteration, imagery, diction.

Too many students take refuge in vagueness (e.g. such phrases as 'a sort of peaceful effect') because they are not in possession of these simple technical terms, and show little experience of the possession of more general terms, to use in discussion. The capacity to use such terms can-not of course be given at second hand—the conditions must be those of first-hand collaboration—in seminars, so that a situation arises, say, over a student's own work, in which the word 'sincere' can be given body for the members of the group. Of course the tutor will have to have examples and exercises at his finger tips: but the important thing is to ensure that the terms are conveyed in meaningful contexts.

In the schoolroom teachers need the most confident possession of such technical and critical terms and the capacity to put them into the simplest language—how else can they begin to discuss poetry and prose with their pupils? To be able to discuss literature in simple terms is necessary with secondary modern children, especially with lower streams, and with all primary school children. The kind of generalisa-tion we have seen in students' exam essays would be Chinese to them. Yet the reading of poetry, short stories and extracts from imaginative fiction is a first requirement in children's English work. To practise it requires of the teacher the capacity to select poems and passages by his own critical acumen, to prepare a lesson on a poem or passage, to introduce the work, and to judge pupils' own writing, both 'critical' and creative. To be able to do this requires a rigorous training in dealing with words at first hand.

* A useful book here is *The Nature of English Poetry* by L. S. Harris, and also, of course, R. F. Brewer's *Art of Versification*. 'Movement' is a term used in a special sense by F. R. Leavis: see *Revaluation*.

16

QUESTIONING FASHION

There is, I believe, a further advantage in working collaboratively, through the seminar. It would help overcome the present difficulty, of making links between the past and the present—especially as the present, in English writing, is so barren of works of positive creative talent, and so full of questionable reputations.

Students should, of course, be encouraged to take an interest in the writing of their own time. But the trouble is that nowadays the English intellectual scene tends to be unduly influenced by the 'higher journalism'. Journalists who write about literary matters may be sincere, but they have to make a living, and there simply cannot be enough genuine new creative work to keep them all in business. Inevitably, they must continually manufacture new fashions, and nowadays the process is closely linked, at a time of high printing costs and dear money, with publishers' promotion campaigns.

The result is that a succession of 'new' writers is boosted, and the most assiduous reader cannot keep up. By the time any serious critic does catch up with any particular work, as likely as not the fashion is ended, and there is no longer any point in publishing a demolition.

The subject is too large to cover fully here. I only note that even colleges of education are susceptible to fashions created by the Sunday newspapers—bright fashions made and essentially established to draw attention to the advertisements on which these papers depend. In one college two girls had adopted *The Poetry of the 1914–1918 War* as the subject for their 'piece of work' because *The Sunday Times* colour supplement published photographs of skulls in French fields—in order to draw the reader's attention to girdle and whisky advertisements. I was doubtful, not only because it suggested too direct an influence from ad.-media cults, but rather more because the mode of choice suggested a lack of offered alternatives.

Not only are books and authors made fashionable by papers, but also associated attitudes to life. And here, I think, there is a dichotomy of which too few people are aware. Seminar work, I would hope, would expose it. That is, while the prevalent fashionable ethos cultivates

futility, 'amorality', a preoccupation with nastiness, as over hard-bitten
sex, hardly any young people naturally share such attitudes, unless they
are mentally ill, anti-social or extremely unhappy.

Yet, though they are full of positive, reparative, constructive, and
altruistic impulses, students are still prepared to write commendations
of fashionable negative postures. Thus, a different kind of insincerity
arises: a division between the genuine self and what one feels expected to
say. The bedevilling influence is literary fashion, linked with publishers'
promotion business, and writers' impulses to become popular and
successful, sometimes by exploiting destructive themes.

So, surely, a course which ends up by concentrating on *Room at the
Top*, *Catcher in the Rye*, *Brighton Rock* and *Lord of the Flies* as 'modern
fiction', at the expense of E. M. Forster, Lawrence and even (say)
George Orwell and Adrian Bell shows itself too much influenced by
fashions. Consider, for instance, a bright college course on Braine, Amis,
etc., in which *Lord of the Flies* was described by the tutor herself as having
'deep and significant theological references...a great novel'. The tutor
seemed surprised to be accused of 'preoccupation with the sordid',
yet she provided no touchstones. Nor did she see ours as a situation in
which, since there was so little apart from journalistic sensationalism
in recent English fiction, it would have perhaps been better to leave
it alone, except for students' own private reading and questions
which they might bring from this. She records that students feel that
any recommended book must be 'good'. Exactly! If a book is
thought to be worth such attention, then a college is endorsing it.
Inevitably, as one knows from talking to students, they then come
to feel that the attitudes to life of the gross and the second rate are
endorsed too often by their English course. (Later they may accept the
—to me shocking—inclusion of Raymond Chandler and Margery
Allingham in the syllabus.)★

Of course, there are aspects which at times make the problem seem
insoluble. For my own part, I cannot feel that any adult of sensitivity,
with a concern for emotional reality, human nature, and creativity,
could really take the novels of Iris Murdoch seriously. Her sexual
situations seem to me manifestations of grotesque hate, devoid of
significance because devoid of substance and emotional truth. Yet her
work is endorsed by academics and threatens to be 'there' in the modern

★ The course referred to above is discussed by H. Morris in *The Use of English*, Spring
1965.

canon. In such a situation, perhaps, the English tutor has to take at least some examples, for discrimination. An interesting example was given in *The Use of English* (Autumn 1963) by Peter Wood, who took some of the novels of Patrick White with sixth-formers, feeling that the valuation of White as 'the Tolstoy of Australian contemporary letters' was exaggerated. The boys set about the task by examining White's style (as one could begin by examining Iris Murdoch's). They found this 'original...in its way...' but 'at times behind this originality may be seen a desire to shock rather than elucidate'. They went on to examine the implications and social attitudes, which they felt were limited:

The failure to present any live opposition to Voss, opposition that the writer does not present merely as contemptible prejudice on the part of those concerned, was a major flaw.

Of his project Peter Wood says:

What was most remarkable...was that these boys had said, in their own way, what was essential about those novels—or had pointed to those features of them from which any open criticism must start. The views were not only expressed trenchantly but were honestly and passionately held.

The boys' attitudes...were...respectful...White's writing was positively 'improving' by contemporary standards...It was the inadequacy of these standards, the irrelevance of the standards of the prevailing literary ethos... that led to their being rejected.

In the background, of course, was the pupils' knowledge of Tolstoy. Where student teachers are concerned, perhaps the best solution is to get them to read Tolstoy, and then let them make their own conclusions about the modern literature they read: 'such writers as Amis and Murdoch' ought then to appear as superficial and destructive, in 'open criticism'.

To discuss style and content in such ways, however, requires a good deal of tough seminar work: and the best subject here, surely, is Lawrence. Even here, however, we are on difficult ground, for Lawrence too has been captured by 'enlightenment'. We can see that, as stories, Lawrence's works are at best vivid and immediate in their rendering of human experience. Elsewhere his implications, if crudely taken as a 'philosophy of sex', are neurotic and false. This is our paradox with him. He is a great artist in our century. But now his false side has been enlisted by fashionable 'enlightenment' for its own destructive ends, and since the *Lady Chatterley* trial some students seem to assume that all bawdiness, and even near-pornography, is not only to be tolerated, but even good for us. Lawrence, because of his unconscious distortions, has become the advance post of a wave of destructive mental preoccupation with sex.

So, even in possessing Lawrence's work, we are engaged in a struggle between love and creativity, and hate and destructiveness—in a way he might have understood (and he would have been horrified to see the effect of his worst influence). The delicate problem is how to challenge, across the age-gap between lecturer and student, the false 'Lawrentian' attitudes taken over by students from fashionable journalism—and worse. As the young man's essay above shows, intuitively he resists these: would not painstaking seminar work bring this out more strongly?

Lawrence is worth fighting over. What I find impossible to endorse is the attention given in college of education English syllabuses to fashionable modern writers, at the expense of works of genuine worth. What I object to is the failure to imply a sense of proportion, as between those books which are having their little moment supported by the promotion campaigns of the periodicals, and books whose value has been patiently established by critical attention.

Where students are spending time on John Braine, Kingsley Amis, Alan Sillitoe, Stan Barstow, J. P. Donleavy and Iris Murdoch at the expense of time given to Lawrence, James Joyce and T. S. Eliot, or even to Hemingway and Scott Fitzgerald, then there seems to me to be an evasion of the essential task, so that things are no better than when students were reading Burke on the Sublime or the Gothick Novel.

Here emerges a problem which cannot be fully dealt with here, though it should be mentioned. In a society of uncertain cultural standards forms of expression have become prevalent—and preferred—which are essentially based on the utterance of an artificial amount of unconscious hate. This hate is not 'contained': little creative energy is directed at accepting and modifying the incorporative and destructive urges. Hate itself is taken as being the means by which to strengthen the identity: if we are 'bad', then we are at least someone. It is acceptable, the implication is, to live at the expense of others: thus what is sometimes offered as 'protest' merely complements the degradation of individuality and subjective life brought about by the philistine ethos of an acquisitive society.

Often the impulse of hate is disguised by 'humour'. Here it is worth invoking psychological comment which links the 'wittiness' of some of those who seek to 'do dirt on life' with their sadism:

The joker by profession...is generally a nervous creature with an unbalanced character who in his jokes really ventilates his insufficiently censored intellectual and moral imperfections, that is, his own infantilisms.*

The witty personality is fundamentally a sadist who needs the laughter of his friends to overcome his social anxieties.†

In literary terms, this becomes the formula that if you make your hero drunk and ridiculous the destructive attack you are making on positives is pardonable.

I am exploring the underlying motives of modern 'hate-writing' elsewhere. What is alarming about the vogue is that it is supported by a promulgation of attitudes which seek, by making themselves out to be 'realistic', to demoralise. Yet, if we study psychoanalytical writing from Suttee to Winnicott, we find such 'realism' is not based on authentic sources of insight, however much it seems to be so. Significantly, apologists for 'frank expression' now increasingly disallow and attack psychoanalysis, perhaps because it is capable of exposing their underlying motives and lack of realism.

There is also a deeper problem in that the solutions of hate to problems of identity, even in expression, are inimical to genuine creativity. To endorse some modern writing seems actually to divert creative interests and promote inhibition and the sense of futility. Our problem is that we cannot get far enough away from the current literary ethos to see how destructive it is. (The scene looks different from abroad, so that a French critic can pronounce some of our 'leading' writers 'nasty schoolboys'.)

In such a situation, drawing students' attention to new work without accepting the promoted falsities is a delicate task. The way to deal with the situation, however, is surely by establishing with students a genuine appreciation of (say) Conrad, E. M. Forster, and Lawrence, with careful discrimination, and then putting a few modern works to the test of seminar discussion. Wide reading of modern writing can then be left to the students' own spare time.

It seems to me that the prevalence of much fashionable modern writ-

---

* Sandor Ferenczi, *Analysis of Wit and the Comical, Further Contributions to the Theory and Technique of Psychoanalysis*, p. 340.
† *International Journal of Psychoanalysis*, vol. xxxvii, 364.

ing in college English syllabuses reflects an unwillingness on the part of the staff to allow enough opportunity for these values to be challenged in local analysis of works of art in actual face-to-face exchange. I think that often neither the reputations nor the works would stand the close attention of seminar work.

This discussion of the assumption and attitudes implicit in books requires some shaking down from traditional academic approaches. In some colleges I have found a fussy concern with 'style', at the expense of concern for what is said—what students offer. A student will inter-ject, for instance, about a passage in a book '—and what a scene !'—only to be told 'Avoid this kind of expression unless reporting dialogue'. The essay is being looked at rather in terms of 'would this please an examiner?' rather than of 'has she got the point?'. So students become obsessed with 'correctness' and write tedious pieces in mean phrases: their true self retreats. Where working-class girls from a warm-blooded environment have written with vernacular vigour, I have sometimes found them marked down because they did not fit the pattern of re-spectable middle-class expression. In this way college lecturers some-times show they have no sense of how literacy is a capacity deeply related to thought and feeling—and to engagement with life. So when-ever students wrote on social issues they often showed insufficient capacity to employ insights gained from literature in their approach to life. They obviously had not 'had it out', time after time, with their tutors, in relation to the implications of the books they were reading. Their answers on advertisements, on children's work, and on such mat-ters as sexual morality showed confusion, prejudice, and too little ability to think, argue, and examine experience critically. There were, of course, notable exceptions. But I felt much lack of free, informal, and ample verbal discussion between staff and students behind 'English' work. It was good to see so many remarks by tutors on students' folders saying 'See me and we'll discuss this'. But where there is least sense of relevance, students' work was cramped, and marked too much like that of children, as if there could be 'right answers'. When the 'right' answers were by way of repeating the false implications of bad books, the situation looked grotesque !

So one sometimes has the impression reading exam scripts that there was a conflict between the middlebrow taste of the staff, and the stu-dents' more challenging nouses. A student who can write this without any comment from her tutor is being let down:

In a rich vein of ironic humour Kingsley Amis is an example of the writers who have tender sympathy with our human frailties which leads to situations of great comic art. We can laugh at and love the misfit though there are no solutions the writer can give us for his problems. . . The bitter cutting humour of Osborne or Braine does not amuse. They show that man is naturally 'mine own executioner' and reveal (consciously or unconsciously?) the spiritual sterility of modern life.

Are college lecturers so cowed by fashionable evaluations that they are prepared to let such judgements go unchallenged? Is Kingsley Amis genuinely concerned about 'the spiritual sterility of modern life'? How do we reconcile with this his fascination with Ian Fleming? A French critic, referring to Amis's work on Fleming as 'aussi brilliant que vide', calls Amis 'un écrivain populaire, se voulant populaire. . .'—is this not a just critical attitude?

Students show themselves at times capable of challenging the fashions, thank heaven. One girl wrote of *Room at the Top*:

It is simply the full acts of sex which are unsatisfactory in 'Room at the Top', but the trivial things also. These are calculated to make people feel that this is true, this is 'real life' Joe remarks at one stage in the book 'If a girl wants her hand to be held it tightens over yours the moment it's touched. . . Somehow it seemed tremendously important that I should hold her hand.' These small, unimportant details are given a magnitude which has a certain appeal, which is a titillation of feelings. . . This playing with the feelings all the time is what I find unsatisfactory about 'Room at the Top'. . . Joe says at one point, that Alice's words were 'empty and tawdry' and I feel that this is an exact description of the writer's aim. . . it does not enhance one's views of life, give any promise of hope, or even provide a thrilling story. . .

(*Student 35*)

Behind the student's direct and sincere look at Braine lies her meticulous work on a section of *The Waste Land*:

There is so much of seeming worth here where each piece of jewellery seeks to outdo the other, whilst perhaps not one is of especial value. There is great pity in this and perhaps Eliot is trying to show that value lies not in the material things of life which will only bring confusion to the holder.

One becomes more certain of the above sentiment on reading of her 'strange synthetic perfumes'. Something which is 'synthetic' is man made. Again there is no true beauty created by the soul, or even by nature's products found outside the body. These perfumes do indeed 'confuse' and puzzle the mind. This word 'confused' is one of Eliot's ambiguities which are, paradoxically, helpful in conveying the meaning of the word. The perfumes 'drowned the sense in odours'. Man-made things, material things have deadened the senses and so killed any true feeling; the love of externals has killed the soul of the woman who is surrounded by such possessions. The air which 'freshened' from the window was 'troubled'. Nature, thus, is morti-fied, almost, by the strangeness of man-made things... *(Student 35)*

She takes modern expression in her stride: and, I think, she left the staff behind in both her explorations. It is more of this that we want—and what we should, I am sure, learn from their explorations is that students can see through many modern fashions.* One often has a related sense of the college lecturers, because they were still thinking in terms of 'passing examinations' in the old grammar school way, not being able to see how good some of the more direct and honest writing about moral issues in literature was. When presented with passages from *King Lear* a student could write:

Lust and cruelty are demonstrated in the action of the play...In passage *A* we find the truth—that they are horrible realities that Lear discovers beneath appearances. With superb insight, they are identified. We see Lear passing through the stages of blind self-will, oblivious to the real worth of people around him, finally being wrenched out of this state of being by the drive of events until he can see human nature with true clarity...The world of appearances is based on artificial and unreal distinctions, slip them off and one finds what Lear finds in the vision. *(Student 5)*

Another:

...one of the most fundamental questions concerns the nature of man him-self and his position in the scheme of things...

Who is it that can tell me who I am? *(Student 19)*

Yet such insights had been awarded no better marks than those students who reproduced fragments from lectures on literary history.

Thus, when we come to modern literature, the lecturer tends to mark according to the current assumptions. A student who writes thus, off her own bat, is even likely to be marked poorly, because her honesty makes the examiner uncomfortably doubtful:

Many of these widely popular novels are like popular music in that both do not last in the public eye for any length of time. Sillitoe's novel has not the strength of quality and consistency in the writing to maintain its position in the public eye when the shocking and titillating factors have lost their

---

* I had to upgrade both pieces of work as examiner.

novelty... Any value it retains will be purely due to Sillitoe's observance of Nottingham working class culture; its interest will be sociological rather than literary.

The failure to make local analysis the basis for discrimination between attitudes to life becomes most depressing in students' discussions of modern drama. It is not enough to discuss the 'message' of a modern work: we must also discuss the quality of the language used to explore experience.

In this sense, students' discussions of the 'theatre of the absurd'often fail to begin to be relevant. Their approach is cowed by the prevalent reputation of this *genre*. I noticed that some of them, answering questions on *avant garde* theatre, sometimes failed to demonstrate any adequate knowledge of the plays. They seemed merely to exist on the fashionable accounts of them in general. So they were unable to be critical of the sentiments and attitudes expressed—yet these must often have been far from their own attitudes. Had they been able to discuss the language locally on the page, they could not have escaped the obligation to challenge the meaning.

Most surprisingly, they saw the function of 'tragedy' as being to express 'futility'. This sentimental concept should be energetically questioned by the English department, and could not be endorsed by anyone who has read any adequate criticism of tragedy (such as that of Santayana, Leavis, L. C. Knights or Raymond Williams).

The prevalent concept of tragedy is that it is about 'the small man' faced with a 'hostile universe'. 'A picture of the small man looking for something to give some kind of meaning to existence...' Surely one could not conduct seminars for long without asking whether it is true that the point of tragedy is to see life as 'pointless', the universe hostile, humanity having no resources, so that the only appropriate response is 'poor us!'.

It seems astonishing that young people should so admire this cult of hopelessness—an attitude they would certainly reject in themselves. Nor do they see how sentimental are some of the symbolic gestures of futility. There seems no irony in this student's account:

They hang themselves on the tree which is the symbol of the life force.

Some modern plays seem to be very sentimental behind the sophisticated façade. Others (like those of Pinter) seem to make their impact by a sadistic attack on the audience, and to set out with the impulse to resolve the author's identity by the utterance of hate. The symbolism often seems crass. Is desultory oblique café conversation all there is to say about 'ordinary' people? When we examine the idioms, is not the concept of human nature too much imprisoned in it? Quotations here (such as the following) are from students at the end of their course: surely such statements should have been the beginning of a requestioning of the assumptions behind modern drama?

The lack of importance of all the characters is emphasised by their conversation, a technique which Pinter has developed to perfection. It is conversation which you hear in dozens of cafés, the conversation of the small people which says nothing because they have few experiences to talk about. It revolves in circles disjointed and inarticulate and again emphasizes the timelessness of the situation because it doesn't get anywhere and says nothing...emphasizes the purposelessness of the small man...

This is the final fear of the small man. He is not strong enough, nor has he the desire to wage war against the external forces of terror and eternity.

It is disturbing to find young teachers failing to challenge this modish cult of the exploitation of sentimental despair. Do students really ask themselves 'Is it worth going on living?'?

It is this denial of progression which makes the world seem timeless and therefore terrifying because there is no purpose in life and begins to ask the question, 'Why go on living?' or 'Why was life given in the first place?'... people watching the play become very small and helpless...life is a repetition of small and rather petty instances...how petty and unimportant his repetitive life is...   (Student 47)

The conclusion is that *Waiting for Godot* ' is a tragedy because it shows all men living futile lives'. Unless such a view is challenged, how can she make sense of *King Lear, Macbeth, Oedipus Rex* or *The Women of Trachis*?

At times, revealingly, students write on these plays in the conventional language of Sunday newspaper fashion:

The cross-talk of vaudeville dialogue is miraculously transformed into pure poetry...   (Student 48)

—'miraculously', perhaps by the sharper critics! The culmination of this preoccupation with the cult of not-art comes when we are told that the characters of Beckett reach the same condition as Birkin in *Women in Love*! At best Birkin embodies the most positive and subtle concepts of manhood in relationship created by the century's greatest imaginative writer: it is sad that a student can find him comparable in any way with the creatures in Beckett's depressive vision of futility, with its abject recommendation of withdrawal from engagement with life and its quality of being 'remarkable nothing':

These are men stripped to the bare bones of self. This is the frightening thing...One then feels entire in oneself, much as Birkin did in *Women in Love*.

One is grateful for the opportunity to disagree with these students' views: but there is not enough sign of their satisfaction with fashionable criticism being challenged. The world of fashionable 'protest' writing is often no real advance on the world of the old West End play and the best-seller. Real discussion ought to expose it.

We may compare the experience of much modern writing with that of *The Ancient Mariner*. In this work 'good' elements are essentially part of the concept of human reality and the work is a striving for wholeness. By contrast a totally negative picture of reality is given by *Waiting for Godot*. There the denial of the creative and constructive elements in human experience is sentimental, since it ignores the evident reality of love and achievement: what is recommended is withdrawal and prostration. This is the obverse of Victorian melodrama, which seeks to deny the harmfulness of hate, or to pretend it can be overcome by magic (sudden conversion, and such). Beckett equally pretends that there is no love—and so no possibility of overcoming hate and inanition: which is not true. A greater truth is declared in the triumphant elation we feel (say) at the end of *Lear*, or in the superb transcending of the threat of annihilation achieved in Gustav Mahler's Ninth Symphony.

So in terms of truth about human reality there seems to me little to choose between Beckett's ultra-depressive pessimism and the bogus optimism of Little Lord Fauntleroy. Life cannot validly be represented as either totally devoid of destructive aggression, nor totally devoid of the good, altruistic, satisfying and meaningful. Both *Godot* and *Peter Pan* are travesties of human truth. The difference is that *Peter Pan* disguises hate and aggression under a faery surface, whereas Beckett makes his aggressive attack directly at the audience, seeking to involve them in inanition. Both seek to involve the audience in essentially regressive attitudes to—or withdrawals from—experience: hardly a new realism.

The assent students give to these works contrasts strikingly with their attitude to such a great artist (who was also a popular writer) as Dickens. Either they are not interested, or their attitude to Dickens is too conventional—he was merely a 'social reformer'. Dickens was more than a 'social reformer' if one accepts that he is dealing with inward conflict in a creative and symbolic way. He explores deeper motivations, and universal human capacities to exert altruism against hate, and love against evil and wrong. In this he contributes to the resolution of identity, and to that sympathy which is on the side of creativity and peace.

This, in an age of not-art and the exploitation of the depressive attitudes in *avant-garde* writing, makes Dickens 'old-fashioned' of course: as one student says:

The people he writes about, and the style in which he writes is now out-of-date, and has a limited appeal...The modern writers of today have just as much sincerity, and they have a message, even if it is only a comment on society, as in *Waiting for Godot* by Samuel Beckett.

Strange reversal, where one of the greatest English prose artists—engaged in a genuine contest with the realities of the subjective world—can be discussed in favour of a minor writer who merely recommends the forfeiture of adult engagement with life altogether.

Dickens's art—its humanity and belief in good and the continuity of life—transcends the confines of his period, as do Chaucer's and Shakespeare's. Dickens strives with great courage to come to terms with human savagery, and to find man 'good' (despite some tormented awareness of his own badness). The situation is the more sad because Dickens is such an excellent writer for use in secondary modern schools—not least because he was a great popular artist.

One result of fashionable pessimism and journalistic 'enlightenment' is widespread confusion in many students' minds about creative and moral issues. Genuine seminar work might have helped clarify these matters. In examining and visiting colleges I have often been distressed by the callowness and inadequacy of the students' moral awareness, and their poor capacity to think about moral issues in relation to literature. This seems to me a sad state of affairs—when the wisdom embodied in fine literature is there to be possessed, if the difficulties of possessing it can be shared.

Sometimes the students showed intuitive good sense, e.g. one girl wrote of *Tess*, 'It seems incredible that a young girl brought up in the country should know so little about life.' But she did not challenge Hardy's obsession with Tess's 'purity' or his willed 'punishment' (under the guise of 'tragedy') of his protagonist. With modern novels students did sometimes compare literary experience with their own experience, often with some good sense: but they were often cowed by fashionable estimates. Intuitive good sense was not sufficiently reinforced by contact with literary sources: students often show too few signs of having been made to think carefully about moral issues, and the light books can cast on them.

Statements such as these from students' work would surely be good starting points for such discussions:

So we learn that, through fate and circumstances Tess loses her virginity but today it is not considered that purity consists only of virginity, today Tess's character would tell of her purity.

To have had some sexual experience before marriage is no longer shocking to us. This is a symbol of how times are changed.

It must be admitted, it [*Room at the Top*] is popular for the sex it contains but other books are better reading for sheer pornography.

Gerald has no emotion and that is why the love-making between he and Gudrun fails, because it is purely sensual. He is never truly satisfied in love or living at any level and dies broken fittingly in the snow-covered Alps.

Students of literature who have the confidence of their tutors, and confidence in their own good sense, should surely do better than this?

Answers by students to questions on Shakespeare's attitudes to human nature and love show how the destructive attitudes of the prevalent literary ethos prevent them from being able to respond to the greatest literature. One student could write that Miranda (who was apparently a 'portrait of Shakespeare's own grand-daughter') was 'the type of naïve, pure, young heroine that Shakespeare liked' and was 'really too good to be true'. What some moderns write looks so 'realistic'—and yet the posture really hides ignorance. Shakespeare made the most profound exploration of lust, sexual reality, and the treacheries of the emotional life. He was 'our contemporary' in that he saw the worst that man was capable of, and at times felt utter despair ('destruction, fang mankind!'). Yet he saw more—and saw the reality of love and creativity in man, too.

So in his late maturity he found he could, without destructive envy, believe in the creative continuity of youth, love, and 'goodness'. Students tend to take this hopefulness as 'unreal'. It would be very sad, if one did not know that they are conforming, largely to hide a lack of inward confidence and critical resources, by which to cry 'Nonsense!' to the pretenders of fashion. It is these potentialities we must foster.

17

WHAT IS A SEMINAR?

Many people speak of 'seminars', but it is often obvious that they are really using the word for an informal lecture, or a general wandering discussion. Sometimes it is taken to mean a discussion with a number of tutors. I use it to mean a discussion, guided by one tutor, which is a real collaboration.

I once heard a distinguished man address a group of tutors in adult education. He spoke of the need for 'exchange' between minds— 'collaboration': and then talked for three-quarters of an hour. He then paused, and apologised, and said he very much felt the occasion should be a 'real exchange'. After a pause, one lady plucked up enough courage to ask if he did not think he had influenced his students too much —and the speaker then talked for another three-quarters of an hour. At the end he thanked us for our 'collaboration', and quite possibly thought that was what it had been.

A seminar is a guided discussion. It begins with the students' experience, and moves according to the pace of their capacity to articulate it. Such a pace allows terms to be established between the members of the group as they go. The tutor's function is to bring discussion back from irrelevancies, and to make clear the values and concepts the group is developing and establishing between them.

The tutor of a seminar must have patience. And he must be able to bear the pains (or at least the apprehension) of various aspects of his method of work. He must, first of all, be able to bear silence, and not feel that he is not earning his keep unless he is talking. He must be able to put a work in front of his students, and to wait for them to talk about it—while he is there waiting to *receive* their utterances. He must be prepared for the whole thing to fail—for no one to say anything, or for no one to say anything relevant—though this does not really happen as one fears it will (students will pardon you for a bad day).

Of course, the tutor will do his best to head off irrelevancies and to stop some students disrupting the whole set-up. But he must be prepared to tolerate a wandering tentative discussion that is making its way towards a conclusion which he (perhaps) desperately wants it to reach, or foresees with ease. It seems maddening to have to wait until the students reach it: will they miss it? But he must wait—and only step in if the point is by-passed, to the general loss.

Then, from time to time, he commits himself more fully. There come moments when by a certain restlessness students betray their exhaustion —that they have had enough of trying to work things out for themselves. They really want you to be yourself, not an expressionless neutral guide. So, the tutor must enter into the discussion, and (virtually) give a lecture. But he works now on the ground established between students and tutor—words fully flavoured, ambiguities searched out, terms agreed upon. Now the students are ready to pounce on him: there's little he can get away with, for the evidence is in everyone's possession.

Such aspects of seminar work explain why it is so much more exhausting than lecturing. Another aspect is that one is often moved, excited, angered, or satisfied—and this process of painful change in one's feelings makes such work far more taxing than giving a 'set piece' lecture.

The greatest value of a seminar is that it conveys that there is no 'right answer'. The point is not to convey 'information' or 'facts' or 'agreement' to the students—the aim is to bring them to argue about critical matters and critical judgements in their own words. They are being asked to work, like responsible independent adults.

This involves change, and anything achieved will be at the actual level of the students' literacy. Developments and conclusions will have emerged from their sensibility and intelligence, and will not be false ones imposed on them.

The exchange is informal, and some of the best progress is made when students attack one another 'across the floor', sometimes in heat. Through such work, students come to *care* about poetry, its creation, its interpretation—and about language in education. After such exchanges they should be better able to take this fresh enthusiastic 'caring' in their own discussion in the classroom.

The tutor, also, by the end of a seminar, may have learned a good deal, too.

These are verbatim extracts from the recording of a seminar conducted by the present writer.*

TUTOR. The question I really want to raise with you to see whether you can answer it is one that came up last week...the two things when you are

* Seminar held at King's College, Cambridge, in October 1962, 2–4.30 p.m., with Education Year students, mostly graduates.

dealing with creative imaginative work...you ask yourself, first of all—
'what do I say to myself about this piece of writing?' And secondly, 'what
do I say to the child?' And thirdly, possibly—if we can get on to this—
what do you do to reinforce the experience you're dealing with? That
takes you on to literature...You look through thirty ballads, say, in the
evening—in your digs in Swindon. You've then got to say, 'which of these
ballads are the best?' Now, this already is a literary critical problem...
you've got to go back to school next day and you've got to say, 'I think
these are the five best ballads. I'm going to read them out and I'm going
to say why I think they're good.' You've then to talk to the children about
why they're good. Let us then take this child's poem and see what we
say about it, in those different ways.*

[The poem is printed above, p. 131.]

Can I leave that for general discussion under those headings?

(*Thirty-second pause*)

GIRL. Should it in fact be 'we'll be out'?

TUTOR. That's how she wrote it.

MAN. What do you think it is in fact...? Do you think that here the girl is
sort of to some extent identifying herself with the lilac? 'Soon my little lilac
tree we'll be out'—as if the girl will 'come out' as well...In other words
she wishes to be an 'angel' herself.

TUTOR. What do others think about this?

(*Confused murmur of dissent*)

TUTOR. One at a time, please. Let's think about this. Here you are in your
digs in Swindon. Here is this poem in an exercise book. What do you
say to yourself about it?

GIRL. That she has *seen* the lilac tree. That she has in fact watched it.

GIRL. What's she trying to say by 'stood over the gate'?

MAN. She's trying to say it stands inside the gate, I s'd think.

MAN. Bigger than the gate!

GIRL. Yes.

GIRL. Probably leaning over the gate.

GIRL. Seems to be a definite identification in her own mind between...
connection between herself and the lilac tree...

MAN. Between her and the angel, too, in fact...she wants to be an angel.

GIRL. Why? An angel in white the best to be seen...

MAN. I think you're misreading that spelling mistake—Couldn't it just be
'will be out'?

GIRL. ...Yes.

* This discussion took place before any of the students had read any discussion of it by me in any
publication.

MAN. Forget that then. She's just describing a tree which is blossoming. She isn't talking about herself at all.

MAN. Yet the angel is attractive—it is after all the best angel to be seen...

GIRL. It's a figure of speech used by parents—'we wouldn't like to go out today, would we?'

MAN. It would seem a highly unlikely spelling mistake, for a child to remember to put in an apostrophe.

GIRL. Yes, she's speaking to it like she speaks to something living, or as she speaks to a doll.

GIRL. She's speaking to it as if to a living thing.

MAN. That's the story. It's not dead...

MAN. I think you're misreading that spelling mistake actually, otherwise why doesn't she bring 'we' in again at the end?

GIRL. It's unconscious...

MAN. It's probably unconscious.

MAN. Oh, don't let's say poetry's unconscious, for goodness' sake—you'll say it's inspiration next!

MAN. Why shouldn't it be?

MAN. Excuse me—it just doesn't happen that way—when a piece gets polished like this. I mean I don't think this is a first attempt. The chances are it's been polished just as if it's been rewritten out again with a few words changed. Therefore you can assume it's not just a subconscious blurting out. Some of it's been deliberately written down, probably as an exercise in a class.

TUTOR. I ought perhaps to explain the conditions. It was in fact written in an examination.* So that there wasn't any chance to 'polish' it.

MAN. What was she asked to write?

TUTOR. She was asked to write a poem about a lilac tree.

MAN. Which she has written...with feeling and sincerity?

MAN. Yes, I wouldn't say there's any attempt to 'please teacher' here.

MAN. The lilac tree is in her experience—and the crown worn by the angel... possibly not.

MAN. I wonder where that came from?

(*Pause*)

TUTOR. I wonder whether we can pursue two things which have come up. First of all this line,

Soon my little lilac tree we'll be out.

It is very interesting, I think, that people pause on this.

The other one about this question of 'poetry being unconscious'—can we ponder that for a bit? Perhaps we can pursue the question of poetry being unconscious first?

---

* An end-of-term class test in fact.

Would anybody like to put their spoke in on this? Somebody has said, 'You'll be calling it inspiration next.'

MAN. Well, for instance, the line

Each day it gets whiter and whiter brighter and brighter.

We might think of this as being the inspiration of an advertisement, on T.V. I mean, it sounds rather like a detergent advertisement, in fact.

MAN. It probably is, isn't it?

MAN. Yes, it probably is but in the context of the poem it's quite delightful.

MAN. It is, yes. Absolutely.

GIRL. Surely the very noticeable thing about that line is that if we were writing it, we couldn't possibly write that because we would be so conscious of it sounding like a detergent advertisement.

MAN. Nor could we refer to angels like that.

GIRL. No. But you have to be writing to some extent unselfconsciously, by inspiration, to produce a line like that—and not noticing, I mean.

MAN. The interesting thing about this inspiration is that she has been asked to write a poem about a lilac tree and she chooses to write about a lilac tree which is *coming out into leaf,* and is also 'standing by a gate'.

MAN. Probably that's usually the place where lilac trees grow.

GIRL. Probably by a gate because she has observed one like that.

GIRL. Probably it's the one in her front garden.

MAN. It also has its roots in the earth, which she had probably also noticed. But that doesn't happen to be in the poem. She has selected what she's interested in.

TUTOR. Why? Why is it a lilac tree that's coming out into leaf and standing over a gate and so on?

MAN. She just likes it. But I don't suppose she's given it any thought and I think that's where this tricky word 'inspiration' or 'unconscious' would come in, you write about what you've noticed and what you feel about without knowing why you've noticed it or why you feel about it...

GIRL. This word 'select'...'she selected what she's interested in'...suggests to me a very deliberate process of 'making' a poem...which is I should have thought not the way a child works...

MAN. Yes, but this in fact has been made—we've been told it's under exam conditions, therefore the child has been given time to compose a poem. This has been 'made' in that sense.

GIRL. I don't think that that has anything to do with it.

TUTOR. As a matter of fact perhaps I can chime in here—to say that obviously something *has* been selected, because coming to school the child has seen a dustcart and elm trees, dogs fighting and lilac trees. It's true she's been told to write about a lilac tree, but even so she has probably seen a number of lilac trees—she happens to select *some.* She selects a particular lilac tree at a particular stage of its growth—coming into blossom.

MAN. What time of year did she write it?

TUTOR. This I don't know. Probably it was about the time when there were lilac trees in blossom. But in fact you can say that children in their writing do select certain things, inevitably, don't they?

GIRL. Oh yes.

MAN. That's what writing always is...it might be conscious and deliberate, but it might just happen accidentally. It's probably that children are half-conscious of what they're really interested in.

MAN. Yes exactly—but a little girl of twelve is 'a little green flower not properly out', isn't she? To put it crudely.

MAN. (scornfully) Oh!

GIRL. She's like a child—becoming into womanhood.

MAN. You're not trying to read that into this poem, are you?

GIRL. Well, I think there's so much that...

MAN. One can't help it...yet it's only eight or nine lines, after all.

MAN. She'd just been told to write about a lilac tree. I don't know if you know anything about lilac trees, but the only time they have any beauty whatever is in the spring when the blossom grows on them. The rest of the time they're really an untidy mess. Therefore for anybody that's been asked to write about a lilac tree the spring is the time to do it. Or as far as I'm concerned. I can't see this idea that she's identifying herself with the lilac tree—that she's about to blossom. I'm not sure about this, but children aged twelve aren't conscious that they're about to blossom!

MAN. No, quite.

GIRL. We didn't say 'conscious'.

MAN. You suggest it's the subconscious coming out?

MAN. Yes!

MAN. So children aged twelve have a subconscious that they're about to blossom?

TUTOR. Do you really think that little girls of twelve aren't conscious that they're going to blossom?

GIRL. They're very interested in the fact of growing at that stage, aren't they?

MAN. I wouldn't say they see it in this form, blossoming out, quite so much. They may look forward to growing up. But I don't think they have this conscious feeling that they will blossom out.

MAN. Of course not. That's what we're all saying.

MAN. They don't think of themselves as becoming more beautiful and more mature and such like...

MAN. I think the less we know about who wrote this poem and when and under what circumstances the better. What isn't important is what's happening in the psychological thing in this that and the other. What's important is what's *here*.

MAN. But we're considering it as the effort of a child. We haven't been given it as just a peice of paper with a poem on it, by Dylan Thomas, or as if a mature person had written it—because I can see very little here that suggests that a child had written it.

GIRL. Surely we're rather messing up the discussion by making presuppositions that the writing of a child is essentially different from the writing of an adult person—as if their relation to the thing they're writing about is essentially different—that it's sort of 'psychological' if you're a child, but 'literary' when you're an adult.

MAN. The adult couldn't have written that whiter and whiter, brighter and brighter, nor have written about an angel in that way. I'm sure this is written by a child.

The discussion had become a little ragged. The tutor felt that it was important to get the students to accept that a child can make metaphorical statements, directly from unconscious phantasy, which she could not articulate explicitly.

TUTOR. I'll read you another poem written by another little girl of twelve in the same examination.

'The Lilac Tree'

How pretty is the lilac tree!
How full of pride she seems to be,
Her purple flowers for to show.
She only does it 'cause she knows
How beautiful she is!
How beautiful she is!

Her leaves all dressed in the prettiest green,
The best of leaves I've ever seen:
She only does it 'cause she knows
How more and more her splendour grows!
How beautiful she is!
How beautiful she is!
And more and more her splendour grows:
How beautiful she is!

But one day, up a gale comes,
The wind it howls, and the sky hums:
They steal the pretty little flowers.
And now, my dears, there's only the boughs:
How ugly she is now!
How ugly she is now!
And now, my dears, there's only the boughs—
How ugly she is now!

I read this as relevant to this matter of 'identification'. This one is surely entirely a poem about the lilac tree as a symbol of 'states of soul', isn't it?

MAN. Yes.

TUTOR. Pursue this a little further. You see, someone here is resisting other people's contentions that this poem is about herself identified with or symbolised by the lilac tree.

MAN. I'll join in resisting, because I think this is unnecessary extra interpretation which is being brought in as to whether the child is identifying with the lilac tree. . .You're introducing an unnecessary concept for interpreting the poem.

GIRL. It's useful to see the close relationship—'Soon my little lilac tree'—the way in which she writes about it, and is seeing some close relationship. . . Most children start this by playing with animals or dolls and they look at a tree or a flower as something they can identify themselves with because it's no contradiction, there's not going to be any fight, because the child can part with the 'ownership' at any time he likes or she likes. Without challenge. And you can be supreme in that sort of world.

TUTOR. Yes.

MAN. Don't you think that if she'd meant to identify herself with the tree, she'd have given us a clue in the next line—she says, my little lilac tree: if she'd meant, we'll be out, she'd have said

Each day you will get whiter and whiter

I'd have thought the difference between the two is while if it is 'we'll'—'it's'—and it's 'it' which is continually used for the lilac tree—suggests that in fact it should be 'will' not 'we'll'. You can't have this identification thing. As with the other poem—people do consider trees grow up and die—but there's no need to say they're considering them in relation to human life.

MAN. I think we're being pedantic about a point of grammar in fact.

MAN. Oh, it's *important*—it's what we're arguing about—whether she's identifying herself with the tree or not. The thing that brings us along this tangent is that 'we'll'.

MAN. I think we'd have been on that 'tangent' even if it had been spelt 'will'.

MAN. No.

MAN. Oh yes.

MAN. That was the first thing mentioned in connection with this—this 'we'll'.

MAN. Yes, yes—it's the most obvious clue. But even without it, one's mind surely moves that way?

MAN. Why can't we just consider this tree as a tree? In the other poem she's not saying we start young and grow old—she's just saying we see these things in nature. . .

MAN. But implicitly it's there. . .

MAN. If you like to think it is, it is.

(*Everybody speaking at once here*)

Children like to be frank about it and they say, 'When I grow up I'll look like an old tree!'

MAN. Not as crude as that, surely?*

GIRL. No. I think that children use metaphors more than similes... As a matter of fact they say they *are* something or something *is* something more than that it is '*like*' it.

TUTOR. Take a poem like Blake's *Sick Rose*:

> *Oh rose, thou art sick*
> *The invisible worm*
> *That flies in the night*
> *In the howling storm*
>
> *Has found out thy bed*
> *Of crimson joy*
> *And his dark secret love*
> *Does thy life destroy.*

Is this an observation about a plant, or a kind of horticultural comment? Or say,

> *Oh, sunflower, weary of time,*
> *That countest the steps of the sun...?*

This seems to me relevant because children respond directly to Blake's *Songs of Innocence and Experience*. Is it possible, then, if you get a poem like this second one—

> *How pretty is the lilac tree!*
> *How full of pride she seems to be?*

As soon as you get that, or as soon as you get 'the lilac tree stood over the gate'—certainly if you address it as 'Soon *my* little lilac tree *we*'ll be out'—surely something other is being done than merely describing a natural object which has struck the writer? What was Blake doing in writing *The Sick Rose* or *The Sunflower*?

MAN. Poor old Blake was obsessed with the whole business anyway.

MAN. Business of what?

MAN. Innocence and experience.

MAN. How might you infer from Blake to this child's poem that there's the tendency to pick out permanent themes of some kind?

TUTOR. I'm saying they're doing the same things in different ways.

MAN. We're at liberty to disagree?

---

* By this point the student who is seeking to protest against the metaphorical element in the child's poem is dominating the discussion and the others are finding him a little obtuse. It is obviously better for him to be cooled off thus by other students than by the Tutor.

MAN. Oh, I think you can. We know the Blake poem is a 'good' poem. Those poems we listened to are good poems, though they're not as good as Blake's. We know that these haven't Blake's talent...

GIRL. We also know that Blake was an adult and deliberately using childish metaphors and language...

MAN. I don't think they *are* childish in Blake.

TUTOR. What do you mean by... It all centres round these words 'deliberate', 'unconscious', 'selecting'. It's rather like the remark somebody made about the poem I read you the other week, *One day a man killed me*, which I said was an Oedipus poem and last year somebody got very hot under the collar and said, 'Hadn't that boy been reading Freud in a Penguin?' There seems a difficulty to believe that a child can unconsciously do what Blake was doing—although Blake was consciously writing as an artist—do you suppose in fact that a line like:

And his dark secret love

or

Has found out thy bed
Of crimson joy

—that those lines are composed by a process of conscious selection? They're really created by 'inspiration', aren't they? What we're really asking is 'Do we believe in inspiration or not?'. I do, you see.

MAN. What exactly do you mean by it, though? Sitting in the bath?

TUTOR. I mean the same thing that happens five times a night to everybody when they are dreaming—the same thing that happens when one is pondering the nature of an experience which you cannot make explicit to yourself. I think before we can go very far in looking at children's work and seeing what value it has we've got to accept that there is a creative process in the mind which is only partially conscious and on which you can't really exert your will, beyond, as it were, being 'midwife' to it.

MAN. True. But we do know from Blake's manuscripts that he went over and over, altering things.

TUTOR. Yes.

MAN. This doesn't preclude some sort of conscious...

TUTOR. No, not at all. One doesn't want to go back to the wild picture of inspiration, you know, like that statue in University College, Oxford— behind bars it is actually. There is a marble statue of Shelley with no clothes on falling upon the thorns of life and bleeding. Or that of Swinburne (wasn't it?) who used to wrap himself up in the hearth-rug and steam would come out of the ends... I think we tend to distrust this picture of inspiration because critics like Eliot have insisted that in order to achieve creation there has to be discipline and application. And so there has to be with children—one needs to say, 'we will sit down and we will work at a

poem. You will do this in examination conditions, and you will be silent.'
But mustn't we also accept the intractability of the creative process?
Obviously something of this kind goes on in all of us, so that any ordinary
person dreams often in the course of every night, and that these dreams
can be interpreted in a great many ways. But certainly these images are
'at work' 'trying to explain' the meaning of some of the experiences, and
trying to make sense of them? I believe some psychologists believe we
can't go on at all without dreams.

MAN. You can do without writing poems, though.

TUTOR. Well, *can* you? Can you really live without some kind of creative
process?

MAN. This girl would never have written this unless you had plonked her
down in the examination room?

MAN. We don't know.

TUTOR. No. But there's the Opies' collections somewhere in the room—
thousands of nursery rhymes, game rhymes, poems of childhood, dealing
with love and sexuality and death in unconscious terms, which really
children preserve themselves because they *need* this sort of culture. Is it
possible to grow up without these creative activities, without nursery
rhymes, without game rhymes, without poetry and without phantasy,
without painting, ideograms, without play-acting, without playing the
ceremonial kind of games of childhood? I maintain that it *isn't*. In fact,
you can't develop your effective powers as a child unless you do these
things—and that the root of them is in unconscious phantasy.

MAN. Yes, I would agree with you entirely. But dreams aren't poems, are
they?

TUTOR. No, no. The whole relationship between the dream and the conscious
activity is very difficult and mysterious ground. I don't want to suggest
there's any direct and simple relationship between them. All I really do
want to say is that—what I would like you to accept, shall we say?—is that
children, although they are consciously writing things down on paper,
can draw on unconscious material which is sometimes so remarkable that
it staggers one—because it means they have perceptions which one
wouldn't credit them with. Here we have one of them—this little girl is in
fact writing about herself in adolescence, knowing that she is going through
a stage of leaving childhood, developing physically into a woman, trying
to preserve a self through this ungainly process:

Heavily laden it sways this way and that.

Through all that she's seeking to keep a sense of self-respect and a sense of
personal beauty and high value—ego ideals. She's reassuring herself by
saying 'Soon, my little lilac-tree, we'll be out'—deliberately using the kind
of speech used by Mummy, the reassuring rhythms of parental speech.

And keeping her eye, if you like, on the blossoming womanhood that she is making her way towards.

Now consciously Florence couldn't write you an essay on *How I get Through Adolescence and What I Feel About It*. But 'poetically' she's very much aware of this process. Just before she wrote this she came to me and said, 'I'm writing stories about teddy-bears and dolls. I don't want to go on doing this. I want...I don't know what I want to write. I want to write *different* stories.' She's very much aware that she's no longer a child. And so she went on to write about a housewife having a row with the sweep. One can see how this poem comes from this process of self-awareness.

...Now, of course, we're faced with the question of *how good a poem* is it? And, so, what to do with it in class. Inevitably you will imply by what you take that certain things are better than others. When you pick up children's books you will select six to read, because you can't read the whole thirty. So what I'm trying to ask is, 'Why does one pick a poem like Florence's?'

I picked it because it *moved* me.

It made me feel here was a child really writing out of the deep feelings in her experience, out of the rhythms that she feels, right deep down, almost physically—'Soon my little lilac tree we'll be out'—that it has this rhythm of reassuring herself. She's caught in this poem an aspect of her experience, expressed it sincerely and beautifully, and this makes me feel it is a good poem. Now to go back to his point briefly, I think to say, 'This is a good poem' is the important thing. In your digs you read through the books, and you put your six aside because those are the ones which have given you, shall we say, a lift of the feelings, have stirred you. You say, those are good poems: I work from this.

The only reason for going into psychological considerations, to ask 'What is Florence doing?' is if your geography master comes and says, 'No, I don't think it's a good poem at all', or if the headmaster comes in and says 'I don't think that's very good. She's misspelt all the words. Tell her how to spell " properly", " heavily laden" —laden's not spelt like that'.

MAN. 'Tis in Anglo-Saxon.

TUTOR. Maybe.

MAN. But you're not doing justice to literary criticism, by going so directly into psychology. I can imagine a poem by a girl like Florence about 'coming out' that would be sheer bad poetry.

TUTOR. Yes.

MAN. But this isn't. And it's not a good poem because it comes from deep down inside. It's a good poem and one of the things about it is that it comes from deep down inside.

TUTOR. Yes.

MAN. Can we have a go, at destroying this poem?

MAN. Do.

MAN. I'm not happy about it now because you started with such a warm introduction to it.

MAN. I think this is a charming poem and I don't care at all about Florence's psyche. This is a good poem and of course it's written by a child and of course it's not as good as an adult poem. What's the matter with it?

MAN. Right: *The lilac tree stood over the gate.*

I always think that oak trees *stand*: lilac trees in fact waver or breeze around. In fact perhaps like an angel would—hover...but certainly not 'stand'.

Do you think that's a fair criticism of that first line? Using 'stood', a very hard word, is wrong there? It could *grow* over the gate or if it were *floating* over the gate it would be all right, But merely to be *standing*—it's a very hard word, *stood* is.

MAN. There's movement there.

MAN. There's no movement in 'stood' at all.

MAN. It's very firm.   *(Some hubbub)*

MAN. Is there not something in the very firmness of it? There is this basis of standing in the trunk that makes it quite straight like a person—and then there's the crown.

MAN. Have you ever seen a lilac tree? One thing about a lilac tree is the rapid way in which it divides. As it comes up it immediately splits up. It's the most untidy tree...

TUTOR. When it's in leaf and flower it has a mass. So surely, it could 'stand' over a gate...?

MAN. If you think lilac trees don't you should write a poem about their not doing it.

MAN. I was preparing for the image at the end—angels aren't supposed to stand any way but tower; but perhaps you'll not accept that?

TUTOR. An angel doesn't stand over?

MAN. They never stand over...

TUTOR. Now you're being facetiously destructive.

MAN. They usually have a cloud in which they're sitting or floating.

TUTOR. You're trying to reduce the poem to cliché. Actually one sees many paintings by Stanley Spencer, or medieval paintings or carvings, in which the angel *standing over* something is just what it is doing: that's how you'd describe it.

MAN. I always see saints or St George standing but never angels standing. They're cherubims, they seem to grow out or float on the ceiling.

GIRL. Isn't there a difference between *standing* and *standing over*?

MAN. *Stood* is even a harder word than *standing*.

TUTOR. The effect of *stood* is surely in the first place to humanise the lilac, although it's a dead metaphor. Standing over the gate is what a mother does when she's waiting for a child to come home, or what the father does when he's looking out at the world going past—'The lilac tree stood over the gate'—you see this mass of blossom and foliage standing over the gate like a human figure. Or at least I do.

GIRL. We shouldn't forget a lilac tree looks so much larger to a child than it does to us.

TUTOR. 'Its young leaves' immediately reinforces this sense of the lilac tree as a human figure. 'Its young leaves moved in the breeze.'

MAN. *I* don't see this *young* anyway: I think it's pushing it about...one calls leaves young and tender, because they are in fact young and tender.

MAN. The interesting thing here is that you're approaching the poem in two different ways. You're asking yourself what a lilac tree does—what *your* perception of a lilac tree is. And he, and us, and I suppose Mr Holbrook also, are trying to see whether, in this poem, the child has a consistent conception of the lilac tree, relative to itself, and as having characteristics and a 'feel' about it—which this seems to have. The child...

MAN. The child's got a 'grip' on the lilac tree.

MAN. She has her own understanding of the lilac tree, put over in its own way consistent with itself, throughout.

MAN. You might go and look at a lilac tree again after reading this poem.

MAN. Can I separate myself from Mr Holbrook? I don't think it's a poem about a girl growing up. I think it's about a lilac tree. Nothing that's been said so far makes me think differently about it.

MAN. Well I think it's always worth going over poetry in this hard-headed way, to try to find out about it.

MAN. But you can only do that if it *has* worked on you.

TUTOR. Yes.

MAN. Oh, since it's worked on you, you've had it any way...

MAN. But I'm not just drunk on this poem—I like to go back and appreciate it.

TUTOR. But it moved you?

MAN. Yes.

TUTOR. Yes. I think this is the essential answer, isn't it? That if a poem moves you, then you can see that it moved Florence, that this comes from 'down here' if you like, without going into psychological details—whether you believe (as I do) that this is about growing up, or (as he does) that this is the picture of a lilac tree. Let's go on...having had this debate in our digs in Swindon with ourselves. Because we've argued so much about it we can't but feel this is an interesting poem. You might have found the other twenty-nine from the class by no means as interesting as this. What would you say to Florence?

MAN. 'Go and write some more poems', I think. What would Florence say if I said, 'Yes of course this is about your growing up.' She'd scream!

GIRL. Well, of course, you wouldn't say that anyway!

MAN. Well, what's the point of thinking about it then?

GIRL. Well, surely, anything that's going to help you to appreciate a poem is relevant...this is a fact about literary appreciation...

TUTOR. What I should have done obviously is to have given you all the poems. I should have said, 'Here are all the poems on "A Lilac Tree" by Class 2 A... which of these is the most interesting?' I've short-circuited this, because in fact Florence's was the most interesting poem. Now it seems to me more useful to say, 'Florence's interest in a flowering tree in spring is associated with her own feelings about growing up.' This means to me that I'd better be jolly careful what I say to Florence and what I certainly do *not* say to Florence is that this is about growing up—because this is to make it all too explicit, to show that I'm interested in it for reasons Florence might be very suspicious about. Florence would dry up at once, quite properly, if a teacher begins to say things like that. What does one say to Florence?

MAN. 'Well done, Florence!'

TUTOR. Yes!

MAN. Surely you never say 'Well done' to a child. You say 'This isn't bad....'.

OTHERS. $\begin{cases} \text{Why not?} \\ \text{Oh!} \end{cases}$ \qquad\qquad (*Hubbub*)

MAN. If you say, 'well done', you must also say they can do much, much better.

TUTOR. Well, you try this on a C stream. If you say this you'll have them all very upset...this is what happens in the grammar school, of course— 'B—', 'C—', 'must do better next time'. I'm sorry—I tend to jump on that one. It seems to me something of a disease in our education, not to be sufficiently generous enough to say, 'I think that's marvellous!'

I'm sorry about that digression. Can we go on now to discuss what we would say to Florence?

MAN. We've still got to think, though, of what happens next—we've got to give an idea of the next direction to take.

MAN. I think she'd know far better than I would what was the next direction.

TUTOR. How would you allow her to discover this, then?

MAN. I'd just leave her alone to get on with it.

GIRL. Well, no, because a child thinks round a certain topic and their own thoughts will come out in a way never mind what subject you give them.

MAN. It's ridiculous. Surely you've got to discipline them. This one wouldn't have got anywhere unless someone had said, 'Sit down in the exam room and write a poem about a lilac tree.'

MAN. I'm not sure about that. It's surely a coincidence that such poetry was written in such conditions. Unusual.

MAN. We've got another one that's quite similar to it...and I think better than this one...

GIRL. It merely indicates that the subject is very carefully chosen, doesn't it? The lilac tree.

MAN. Come, come! It could have been a horse-chestnut tree or a sycamore tree—it didn't have to be a lilac tree. Any flowering tree...

MAN. I think it did for this girl. I think you need to do more than merely dish out the poem and say, 'That's very good Florence—press on!'—don't you?

GIRL. You can point out that she has described something well. She has looked at something and has written about it, and not just reproduced something she has read.

MAN. In my experience in a grammar school if you set a subject like this you'd get a whole batch of 'literary 'poems—of the sort they think are likely to please teacher. You'd perhaps only get this non-literary poem in a secondary modern.

MAN. I don't think you can divide your praise. You can't say, 'I like this poem on the whole but there are some things about it I'd like to see improved.' I think to do that would be fatal. And I think that one needs to reinforce the success—the child realises she has been successful and one needs to reinforce that. The child has in the poem itself too realised her own success—the last four lines, for instance, are achieved because there's a sort of impulsive line which seems successful to her—she goes on to another impulse, 'soon it's like a crown—you know, a crown worn by an *angel*.' Then she goes on to another impulse—an almost instinctive impulse —'an angel in white'—then the last impulse—'the best to be seen'.

TUTOR. So she is almost recording the success that the poem is...in the last four lines?

MAN. Yes,—each success leading to a further success.* What interests me more is what do you say to a child who has written a bad poem?

TUTOR. How would you approach it? Here you are with your six good poems and your twenty-four not so good, as you go into your classroom—what do you say?

GIRL. Surely it's important that a child of this sort of age and self-confidence can't really do anything about the poem once it's written? So that whether a poem's good or bad the child inevitably has to start again and to start to

---

* This seems to me a very relevant and pertinent observation of the triumphant elation one may observe in Florence's poem and many children's poems—revealing the sheer joy in successful composition, by rhythmical enactment.

make something that's quite different and in a sense has no connection? If you start trying to improve on something that's already been written. . . .

MAN. I think anything 'adult' in this poem would be false.

TUTOR. I think I should say that I'm against—as far as poems are concerned—getting the child to rewrite or refurbish. Not actually in the heat of creation. Children will come up to you and say 'what's a rhyme for *brown*?' or something and then you can stand there and write twenty rhymes out—and then they go back and write a line that doesn't rhyme anyway. In the heat of creation you've got to be willing to assist—or rather, it isn't so much assist as rather bolster up, to reinforce the process. After it's finished (this is based only on my own experience of secondary modern work) I'm not in favour of re-writing. Your point is a useful one, that having had this creative experience they have as it were 'gone on a stage' in looking at experience, whatever the experience is, whether it's a lilac tree or Florence growing—the triumph that you see in those lines is in that she's reached a stage of perception—she's glad about spring and being a little girl and about lilac trees. The next stage is not necessarily going to be writing a poem about roses, or writing another free verse about a natural object. The next stage will take whatever form Florence's unconscious processes dictate—you have to allow for this and be more interested in the nature of experience she's going for next, which might be very possibly here a romantic story about a youthful love venture, because she's suddenly seen (I would say) in another way how she's going to be a woman. Next stage is to start explaining this in terms of, say, finding a lost baby, or a housewife ordering groceries or some love adventure. This is the next stage in experience—so it isn't a 'literary' process, in which you say, 'She's written that—now she can jolly well write a sonnet.' But being interested in the stages of development, and relating one's choice of form and content—and literature—to this.

MAN. In talking to the class about all the poems, the most interesting question to me is, how would you talk to them about the use of cliché? It seems to make the most impressive thing in this poem, one of the most impressive things was the taking of a cliché and transforming it through sheer sincerity into something pretty grand. The poem you read last week was full of film language, somehow it was transformed. Here the cliché we've commented on, 'it gets whiter and whiter brighter and brighter,' is also transformed. Now presumably most of the poems which have not succeeded to the same extent will have not have succeeded because they have used cliché in the worst sense. How do you put this over?

GIRL. It's not the words that make the cliché.

MAN. Nothing else does.

GIRL. May I say what I did in a case like this?

TUTOR. Do.

GIRL. The children were about twelve, twelvish, secondary modern children. And we had been reading various things about the sea and boats and things like that. And I'd been reading before I'd come into the class a poem about an eagle—a very weak thing. And one day I said to them, if you like to write a poem today on some of these things you can do, some of them had decided to write a play in one corner and the others were writing stories or poems. There was one little chap who wrote a poem on an eagle. Far from being the best in the class he found this certainly a bit above him—you'd put him into a B stream. And he was certainly about half-way down the class. He produced a delightful poem and I duplicated his poem with the other poem which was weak. And I passed these sheets round the class— nobody knew who had written either poem except the little chap himself and I'd warned him beforehand not to say anything. He didn't. And the children all decided they liked his poem much better and of course he was thrilled to bits and they found out for instance why they preferred his poem—because it was more spontaneous, much freer, and so on. And what I was surprised about was that the children were so surprised to discover that it was his poem. And when they discovered it was by a little boy who didn't get on so well it was a spur to all of them to settle down again, and they did, and produced much better stuff than what they had done when there wasn't this spur of competition. I haven't got the poem now, but it was a short poem, four verses, two lines in each.

MAN. A similar question is how do you deal with the mistakes in this poem?

GIRL. In a thing like this it's better not to point them out too much.

MAN. You think so, do you?

GIRL. Yes, you interrupt the creative thing.

MAN. They don't matter.

MAN. No they don't. One has to get it across some time or other, or they go on repeating mistakes again and again.

GIRL. Surely the mistakes in this are phonetic mistakes and if you start to alter them you'll alter the whole use of words?

GIRL. What you'd do is in the next session you have you'd inadvertently put the word 'leaves' up on the board.

GIRL. You can do that by saying, can't you hear it? Say it and just listen once. They'll correct it themselves and this is much easier rather than if you dictate to them and say, 'You don't spell it this way.'

MAN. Obviously it would ruin the poem to say 'rub out leafs and put in leaves'.

GIRL. Yes, you can't do that.

TUTOR. You can keep the word 'angels' for a spelling test at the end of the week. In a sense this is rather a literary poem. Florence is working so close

to her deeper feelings that the language is heightened. In ordinary conversation she wouldn't talk about a thing being 'heavily laden'. She
probably wouldn't talk, though she might, I suppose, though it would be
unusual for her to say 'not properly out'. But 'heavily laden' is certainly
literary—it goes with the rather solemn state in which she writes this poem.
So if you were to pounce on 'probly' and 'heafely ladan' for spelling—
these are the words where she's experimenting with language to get over
an apprehension and if you start falling on those and crossing them out
you're going to knock the actual creative aspiration itself on the head.
And it won't be long, of course, before you get nothing out of Florence
and the others except purely mechanical things in which they're not
interested, in which they don't want to experiment, because they don't
care. Like the little girl I read out last week, using the phrase 'absolutely
horrified'—if you are pernickity you're never going to get anybody saying
'heavily laden' or 'absolutely horrified' or any such extravagant phrases.
A little C-stream girl writing about finding a baby on the step wrote 'at
my disaster'—'wrong', but a new marvellous expression, too. This kind
of panache, adventurousness in language, is just what one wants, and the
way to kill it is by stepping on everybody's spelling. So I'm glad you
raised that.

GIRL. If you have some system of putting poems up on the wall or duplicating
them they'll be quite pleased to have them corrected then.

TUTOR. Yes, or to use them as a 'reader'. The first thing to do with a poem
like this is to give it an audience. And you do this either by writing it out,
or by putting it on the wall. And if you're putting it in a Class Magazine,
of course you correct it. And nobody's more pleased than Florence. It's a
painless way of doing it—it's also the right way of doing it, because she
then *sees* the word 'heavily' and she sees it in the context in which she's
jolly interested—and so it gets fixed in her mind if anything can, if she's
any good at spelling...

There are many points in such an exchange on which to base further
work. Here, obviously, a discussion of some poems by Blake could
usefully follow. We need to explore further the relationship between
sincerity of content, and quality of expression. The students could also
look at some nursery rhymes, and possibly some Shakespeare songs—
to discuss the nature of poetic imagery (e.g. *Fear no more the heat o' the
sun*).

The value of references back and forth between the nature of the
child's apprehension and expression, and literature proper, with student
teachers, will be obvious.

A further—not, I hope, too obvious—point may here be made, about the seminar as an experience of education itself. It implies a degree of equality between all present in searching experience: 'we all share the same darkness'.

Thus, a student discussing a poem in a seminar is also brought to consider his own experience, and to allow a flux between his inward life and that of the poet, in a disciplined situation in which the lecturer seeks to receive from him the products of his effort (and seeks to prevent him avoiding the disturbances of feeling, by irrelevance, facetiousness, or other defence measures). The lecturer and the members of a seminar enjoy, are moved, are upset, discover, gain insight, *together*, in collaboration: the complex is one in which growth may take place. Though such a seminar will be a valuable experience of an educational technique, it will also help the student to grow a little as a whole person at the same time. It can be excitingly relevant (not necessarily directly so) to his personal problems as a young man. It will yield insight into 'possible' as well as past experience. Yet all this occurs in the impersonal context of the discussion of literature—on the 'third ground' of creative exploration of experience as with children in an English class. This itself has implications about the kind of place a college of education should be.

Some writers schematise the problems of a student, as Havinghurst does:*

Achieving new and more mature relations with age-mates of both sexes.
Achieving a masculine or feminine social role.
Accepting one's physique and using the body effectively.
Achieving emotional independence of parents and other adults.
Achieving assurance of economic independence.
Selecting and preparing for an occupation.
Preparing for marriage and family life.
Developing intellectual skills and concepts necessary for civic independence.
Desiring and achieving socially responsible behaviour.
Acquiring a set of values and an ethical system as a guide to behaviour.

Eric Allen, who quotes these in an article, says 'students undertaking professional training are likely to be actively engaged still in some of these tests'. (I like that 'still'—which of us could say we have *begun* to solve any of them?) Such a list is quite useful: but it tends to reduce subtle complexities of living to oversimplifications, and thus to be

* R. J. Havinghurst, *Human Development in Education* (New York, 1953).

superficial (to tend, that is, towards 'parts' of persons again). This list, for instance, omits many aspects of one's relationship with oneself—that need to become assured of one's continuing identity, and of one's capacity to exert a reparative goodness over one's impulses towards destruction and envy, that underlies one's 'ethical system'. The latter in any case is often no more than a rationalisation of an unconscious structure of attitudes: one's actual behaviour is often impelled by unconscious motives, though it is the 'reparative impulse' that tends to make the 'ordinary person' a 'good ordinary person'.

To help young adults with such problems we need to meet them 'in the word' in the fullest sense, in paying attention to aspects of inward and outward reality. They learn thus to 'meet themselves in the word' from this experience. They should be able to go out to meet children in the word too. It is this 'meeting' that this book explores, and the seminar should surely be the chief focus for this meeting. It is natural to move, in a seminar about poetry, into a general discussion of attitudes to life, or moral issues—and in such an impersonal context it is sometimes possible for the deepest personal problems to emerge. But the main values, of course, are the close attention to language required by the circumstances of discussion, and the experience of such collaboration itself.

Seminars of the kind suggested here will only work where there is a feeling of equality between tutors and students. Two practical points are worth making, finally. Where contentious material is being discussed (as with modern works as suggested in the last chapter) it is sometimes useful to have more than one tutor present. And it is also of value to train students to be capable of going off by themselves, without a tutor, to hold a seminar, or for a student to take a seminar from time to time with the tutor present. All these are valuable experiences of education themselves.

SYLLABUS AND RESOURCES

18

THEMES FOR EXPLORATION

Of course, even if we approach teacher training in this dimension, we still need a syllabus on paper. What do we want in practice? Something perhaps rather like what is hoped for in the *Evidence presented to the Plowden Committee* by the National Association for the Teaching of English (*Teacher Training in English*, 1964). Their final conclusion is this:

teachers are unlikely to develop the permissive, imaginative and sympathetic teaching methods we have recommended if they are trained at college by very different methods. In too many colleges, the curriculum involves long hours of lectures and formal teaching, and student social life is still restricted and over-managed. We believe that the colleges should encourage their students to work more on their own and in small groups, in a spirit of enquiry and discovery... (*p. 4, paragraph 83*)

Despite its generalities (I'm not quite sure about the meaning of the word 'permissive'), the heart of this N.A.T.E. evidence is in the right place, emphasising that 'The Training College must provide the kind of environment which will contribute to the emotional and intellectual development of the intending teachers'. But little is said in specific terms of work—the memo resorts to high-sounding phrases:

They must have rich and inspired teaching...they must *at all costs* never lose the spirit of idealism—[what are the costs, one wonders? Does there have to be a cost?]...generous[?] basic course in English...literature should be a worthwhile [?] extension of her experience...giving another dimension (the visual arts, etc.) to the student's understanding of the relationship between experience and imitative expression of which language is the most universal and pervasive form.

However, despite the vagueness of many of its terms (I do not quite understand here the meaning of 'imitative', 'universal' and 'pervasive') this evidence is worth turning to, for its emphasis on the importance of good speech training, the need for more attention to children's literature, the need for combination with the other arts, the need for better libraries in colleges of education, and the need for better probationary supervision during a young teacher's first year and for 'in-service training'.

Here I am concerned to bring discussion down to earth as much as possible, and to try to give substance to such terms as 'generous', 'worthwhile', 'at all costs' and so forth—in terms of what is needed, of course, rather than in terms of 'esteem' or 'status'. I use the latter words advisedly, because the need for improvement here and there seems to be conceived in terms borrowed from the wasteful struggle between the Gas Board and the Electricity Board (e.g. 'I think that if we did so and so we might attract more students who now go to Blank...'). Valid reform needs to be thorough, and based on a full understanding of what are our aims, rather than on questions of local 'status'.

By now I have, by implication and by example, defined the dimension in which we must work if we are to close the gap between what a teacher needs daily in the classroom, and what he is given over his three years' training. Now perhaps we can approach the syllabus with a greater sense of realism. For one thing we must stop dividing 'English' as a subject from English method, and dividing both from the experience of the method of pursuing the subject, as in the seminar. We must read literature at first hand, and cease to divide language from literature. We shall cease demanding the kind of essay students will never do again in their lives—composed from sources 'outside' the work. If we are to have essays they should be of a kind one could use as teaching notes, explorations of true and relevant response.

We are going to abjure the kind of examination and its questions which make false implications about education and how one prepares to take part in it.

Let us question the assumptions behind essays required in examinations at one Department of Education: under psychology, for instance:

'Any subject can be taught effectively in some intellectually honest form to any child at any stage of development' (J. S. Bruner). Discuss this hypothesis in the light of Piaget's account of the psychological development of operations.

No doubt an interesting intellectual question: but a student who answered it well might have no capacity to match his teaching to the needs of his pupils at any one stage, nor does it call on students to discuss any actual experience of their own, of children's stages of development, and the needs that arise during them. Students are being asked to discuss, in the abstract, a complex living process, which in the actual classroom

238

they might have a 'feeling' for, and a gift to deal with—but asked to discuss it as if it were static, and a question of schematised entities, instead of a flux of living in actual, whole, exploring persons. This is the penalty of dividing method and theory from preparation for living experience.

A similar objection seems to me to apply to such a question, under 'Educational Aims, Organisation and Practice':

Develop the theme that, 'a schoolmaster should think *with*, but not *like*, his pupils'.

In fact such a question is almost certainly answered in terms of the belle-lettrist display of affected wisdom. But in demanding an answer to such a question the department is essentially not inviting students to ponder their experience sincerely in the ruthless pursuit of insight—as it might have done had the paper set some children's writing for comment. What this kind of 'information about the teaching situation' approach tends to foster is a 'knowingness' about living processes that tends, if anything, to undermine the intuitive gift for teaching and for confident commitment to the experience of the teaching situation.

So, too, in English, such a department requires an essay answer virtually asking 'Do you now know how to teach English?'—from the outside, in general terms, rather than a demonstration of how to approach, with sensitivity, a particular poem, or piece of writing, in a living context:

'Lack of interest seems one of the main reasons for poor English...the lack of a sense of relevance...seems particularly marked among prospective school-leavers of mediocre ability.' Comment, and suggest means of preventing this boredom, in any type of secondary school with which you are acquainted, or at any particular stage therein.

This is, at first sight, a relevant question: certainly the problem is one of which one has bitter experience. But to answer it needs more experience than most student teachers have, and a wider reference than the question implies: it really requires a novel or full-scale book of reportage to answer. Is there any point in asking such a question? In an hour little more can be done in answer to such a question except to turn out platitudes, and to display superficial knowledge. To do so is no indication that a student could in life overcome boredom in 4 B—indeed, the whole essence of a successful approach to 4 B would be that it was conceived in living terms, rather than in abstract terms encouraged by requiring this kind of essay.

What kind of writing are we going to ask for, then? I suggest among other things:

(*a*) Longer pieces of writing on children, and on presenting English to children, written from experience—at length (10,000–50,000 words).

(*b*) Essays of a literary critical kind on poems and novels or on actual pieces of children's writing.

(*c*) Papers to be presented to seminars, and completed after discussion at the seminar.

(All of these to be done in term, and made the basis of assessment. No examination!)

We also need to bear in mind the need for students to spend more time with children, and to study more of the literature especially written for children: and the need to write about these experiences.

Our time is beginning to fill up!

But let us try to relate the time at our disposal to the procedure for 'meeting in the word'.

We begin from students' own reading. We want them to become able to read in a more attentive way than the way they read in desultory leisure. So, we shall exercise their reading capacities in seminar work. Equipped with this sharpened power of attentive reading, the tutor chooses poems, or short passages of prose, at the beginning, which he knows, to open up with them discussion of relevant concepts, of *how* to discuss literature and in what terms. So, we read some poems by Edward Thomas, or D. H. Lawrence, or Ezra Pound. These poems, we know, will raise general questions of two kinds: of how poetry registers and communicates inward states, and of the nature of man in the modern world. Other books are suggested: (say) Helen Thomas's *World Without End*, Frieda Lawrence's *Not I but the Wind* and E. T.'s *Memoir* of Lawrence, are in the back of the mind—and in the booklist. But in seminar work, discussion is concentrated on *words*—on tone, rhythm, manner, image, nuance, attitude, and the way in which we can profitably discuss these.

The tutor will perhaps introduce *Goodbye to All That* or the poetry of Wilfrid Owen. Or he can go from *Hugh Selwyn Mauberley* to discuss the whole ethos of the acquisitive society (and so on to Lawrence, perhaps).

So we proceed, both by sharpening attention to the words and sharpening awareness of the world we live in. Perhaps the tutor turns to a short story such as Lawrence's *Things,* or the more deeply relevant *Odour of Chrysanthemums.* Here the essential needs for relationship, to find 'the other' in life, are seen to be thwarted by the social, psychic and emotional conditions brought by the industrial revolution, and the way of life it imposed, as in the Nottingham coalfields (here E. P. Thompson's *The Making of the English Working Class* comes to mind, as a useful source of historical understanding of this process). From *Odour of Chrysanthemums* the tutor might go on to *Sons and Lovers* and *Women in Love.* For myself, I might well embark on the territory I explore in *The Quest for Love* around Lawrence's attitudes to the industrial community, and to the potentialities of relationship in the modern world—examining their unconscious origins.

Now I have written half a page about an imaginary beginning to a course in literature, simply following my own interests. But what I have outlined by now is virtually six months' work—or, say, twenty-four meetings of two hours each. At these meetings, for the first six at least, the students would need themselves to talk for an hour at the beginning of each meeting—and would interrupt me a good deal at other times, in the process of getting things straight. The two short stories have been read aloud during the sessions. We would hope for such an attentiveness as one often experiences in adult work—the human atmosphere of a good class in which it is not uncommon to look up and see members in tears. We would hope to be some distance from the atmosphere of a university lecture, which is often a 'turn'—a good deal of showing off, intellectual display and some defence against feeling. In a good literature class no one minds a silent pause after a reading of two or three minutes—for this is a tribute to the artist, and a mark of one's depth of respect in the response, that one does not want to articulate superficially. But to work in such a real way takes time: and we must have no pretences about it.

When all students have completed their reading of each novel, then an hour must be left for the *students* to articulate their experience of it first, before we begin to give our reading (which will take 2–8 hours for each book).

So if we take my imaginary course, we can make out a time-table such as the following:

	hours
Some poems by Ezra Pound	2
Some poems by D. H. Lawrence	2

This is a first term's work! The students will have talked for approximately 10 hours themselves, out of 20 hours' meeting: the tutor himself 10. By the end of the course the tutor can use many terms which he knows that his students understand and can use themselves—and he also knows that they have read all the works discussed, some of them several times. From this firm ground he can deal with the two novels in the 14 weeks left to him, taking up rather more of the time himself. But, when he takes a significant passage (e.g. the burning of the bread in *Sons and Lovers*, or the 'Rabbit' chapter in *Women in Love*) he will read the chapter aloud, and then get students to discuss it at length, in terms of local attention to the language, and response to the poetic symbolism.

Now, this seems to me a realistic way of working, originally based on the approach in adult education, which can be applied to the training of students. It is the way I work in my own seminars. The differences, however, are these:

(1) There is a large gap in age and experience, which makes it difficult for the students, who fear making fools of themselves, by revealing their ignorance of life to an older person. This can be overcome in time, and the whole atmosphere of the instruction, of course, counts. Some university tutors use younger research men to take seminars with first-year men, in order to train them in this method of work, up to the time when the age gap ceases to matter. But I think we should try to 'get through' at once, since there is also a big age gap between teacher and school children too, after all.

(2) The experience of training for examinations at O and A level leaves students with a 'right answer' complex. This becomes combined with their defence against letting their callowness and weakness become exposed—it has a severely uncreative effect, and manifests against honesty, which requires an occasional discovery of one's own inade-

quacies, and of 'useful depression'. Whereas members of an adult class will speak up forthrightly at once, students tend to hesitate, or to pursue irrelevance, in order to put off as long as possible the risk of committing themselves to an opinion which they feel might be found 'wrong'. They are also rather afraid of giving way to feeling. The complement to this is the way undergraduates patronise their examiners—thereby exerting the maximum degree of defence by flannel. This 'right answer' complex is found less among training college students (though these still have to live down O and A level habits), but is a barrier with 'education year' people, who find it alarming to have to accept that there are no certainties in dealing with literacy, literature and imagination—that all is merely the exchange of formulations ('This is so, is it not?', as F. R. Leavis puts it) which, while they admit—invite—discussion, are beyond investigation in terms of any ultimate certainties.

(3) While in adult education critical terms can be established easily with people who are not on the whole acquainted with them, with students these have to be established in the face of bad habits (such as those inculcated for purposes of 'appreciation' in examinations) and of the tendency to display as many technical terms as possible whether or not they are understood etc.—while acquaintance with terms qualifying content (such as 'sentimental') tend to be wildly used, or used in an unbalanced way, unrelated to actual feelings of response. (Cf. 'You are not to think like this before...')

Such a realistic glimpse at what seems to me a realistic way to work (if works are to be read) reveals the impossible task many colleges set themselves in their syllabuses:

In the first two years of the Course, both colleges propose making a general survey of English literature from Chaucer to the immediately contemporary ...the emphasis of study at —— college being on forms and topics and at —— college on periods.   (*Proposed Syllabus for B.Ed.*)

This proposed B.Ed. syllabus, that is, recommends a condensation of the following into two years: here is the present three-year course:

Main English Syllabus: Three-Year Course

General Aim. The development of the student's imagination through the study of selected writers and periods in literary history.

Year I. The growth of traditional forms—poetry, prose and drama; examples from Homer to Eliot. Significant early writers: Dante, Chaucer and others.

The sixteenth century in the writings of Spenser, Lyly, Shakespeare and others.

Year II. Seventeenth-century poetry—Milton, the Metaphysicals; the change in the theatre tradition; the coming of modern prose, in Dryden.

Some writers of the seventeenth and eighteenth centuries; important verse and prose works of the period, notably the novel.

Year III. The romantic period and the Victorian, to about 1880; chosen individual writers from Wordsworth to Hopkins. Later nineteenth- and twentieth-century literature from Whitman to the present day. Tutorial study, emerging from earlier discussions, group and individual, of topics selected by individual students from work done in the lecture course.

The Main English Course is a three-year course for all the students who choose it; Advanced Main students will do more intensive and detailed work on topics selected from the Course as a whole.

The *General Aim* is unexceptional: what is *not possible* in real life is the realisation of the general aim in terms of the outline set out under the three years. Such colleges give perhaps 2 hours a week to 'Main' English students, and another 2 hour session to 'English for all'. Perhaps there will be an additional seminar on children's poetry (under 'method'). But what seems likely is that the syllabus has to be covered in 4 hours per week, or 48 hours each term—taking into account interruptions from teaching practice, and so on, something like 120 hours in a year.

If we take the first year's syllabus, and turn it into terms of actual reading, what do we find? Taking Chaucer and Shakespeare as major figures, we shall surely read *The Prologue*, two Tales and perhaps something from *Troylus and Criseyde*. From Shakespeare, surely, we need a history play, two tragedies, *The Sonnets* and a late 'dark' comedy at least. From Dante we can take the *Vita Nuova* as a minimum, and 'from Homer to Eliot' nothing less than forty poems will suffice, surely, as representing the various 'forms' of poetry. What about 'prose' and 'drama'? Greek drama requires *Agamemnon*, *Antigone*, the *Bacchae*. At least four novels? Other Elizabethan writers?

Now let us do a little sum: to really read these works in classes of twenty, in seminar conditions, will require the following time:

	hours
Dante: *Vita Nuova* (superficially)	8
Forty poems, Homer to Eliot	80
Chaucer's *Prologue* (superficially)	8

Two tales by Chaucer, reading and discussing	8
Troylus and Criseyde	4
Shakespeare's *Sonnets*	16
Henry IV Part 1	8
Macbeth	10
King Lear	12
The Winter's Tale	10
Greek Drama, three plays (superficially)	8
Four novels	
Emma	10
Middlemarch	10
Bleak House	10
Sons and Lovers	8
Other Elizabethans:	
Nashe	4
Marlowe: *Faustus*	8
Tourneur, etc.	10
TOTAL	232
TIME AVAILABLE (day)	120

This very rough sketch of the possibilities reveals that such a syllabus, in any terms of actual shared reading, is unreal: there is only time to read about *half* the works required. And so, what happens is that the works are *lectured about*—and it is *hoped* that the students read them. The effect in the end is that students write a series of essays half from their lecture notes (as the lecturer speeds on his way from Homer to Eliot) and half from the *Pelican Guide to English Literature*, or *The Cambridge History of English Literature*. This is not, alas, a training in reading—and can give no adequate possession of 'signposts'—indications—in literature, since there is too little possession of the works themselves.

The same is true of a college which takes 'periods' instead of 'forms'. These begin with a similar breathless rush from Homer to Eliot in the first year:

In —— College each period covers half a century and is followed for some four weeks, after an introductory week of lectures, the student studying one major work in a year-class and one of his own choice, from a prescribed list, in seminars...

(For the B.Ed. the third and fourth years will be spent in a more intensive examination of areas and writers selected from the general survey undertaken in the first two years.)

Specimen period
  Centre (year-class)   *The Waste Land*
  Lectures (one week)   Background to the Period; *Ulysses*; Strindberg
                       and Ibsen; Yeats; D. H. Lawrence
  Prescribed list for students'   The poems of Hardy; two novels of Henry
  choices (four weeks)   James; *Nostromo*; two novels of D. H. Law-
                       rence; three plays of Bernard Shaw; two plays
                       each of Strindberg and Ibsen; Hopkins; Yeats

Here the degree of unrelatedness of dimension seems grotesque. The student, it seems, studies *The Waste Land* for something like 50 hours throughout the year (it surely needs no more than 8?) while expected to make a choice from Strindberg, Ibsen, Joyce, Yeats and Lawrence after a week's perplexity of being lectured at, on some of the most difficult and complex works of art in English! Four weeks are then spent on *Nostromo* or two novels of Henry James, or two novels of Lawrence's according to the student's own choice. But, presumably, very much on his own, since the tutors could hardly conduct seminars on all those pre-scribed works, with students who wanted to study them, throughout the 4 weeks.

Simply to place alongside each work the amount of time—the mini-mum time—it takes for a proper give-and-take class to get through a work, indicates how any realistic approach would need to take a quite different form:

	hours
Ulysses	20
Nostromo	14
Hardy's poems	14
Hopkins	20
Women in Love	16

The reason for the inadequacy of students' reading and their in-capacity to articulate their experience of reading is directly related to the lack of realism with which tutors estimate in their syllabuses the amount that can actually be covered. My rough calculations have left no room for discussion of 'background', of relevant social problems such as the mass media or popular taste, for discussion of literature for children, or children's writing, and of fancy and imagination! It leaves no room for creative writing, speech work—or 'method'. Where these have been introduced they have probably caused an exacerbation of the tendency to 'cover ground' by lectures which must at times seem to the students like the commentary to an all-inclusive coach tour.

Probably many courses are better than their syllabuses indicate—though it seems unlikely that they are better than a study of the year's essays, pieces of longer work, and examination papers reveals. It would certainly seem that many syllabuses are deceptive—not necessarily window-dressing, but self-deception. With most, their baffling confusion reveals the lack of a sense of relevance: this is the kind of proposal being put up for the new B.Ed.:

In the Second Year, the topics selected would show the continuity of nineteenth-century and twentieth-century literature, and would use as centres or points of departure significant individuals and works: Hopkins, Whitman, Arnold, Hardy, Eliot, Yeats, Thomas, Hughes; Jane Austen (*Emma*), Emily Brontë (*Wuthering Heights*), George Eliot (*Mill on the Floss, Middlemarch*), Lawrence (*Sons and Lovers*), Forster (*Howard's End, Passage to India*); Eliot, Fry, Auden, Bolt, Ionesco. Wordsworth and the Romantics would lead to some study of the great Victorians—Tennyson, Browning, Carlyle, Newman.

In the Third Year, topics would include the poetry of Herbert and Vaughan —related to the work of Milton; the criticism and the verse satire of Dryden; Defoe (*Colonel Jack, Robinson Crusoe*), Bunyan; selected novels of Richardson, Fielding, Sterne (*Pamela, Tom Jones, Tristram Shandy*); the essays and letters of the late seventeenth and eighteenth centuries (Gray, Walpole, Cowper, Addison, Steele, Johnson); *Vathek, Castle of Otranto, Rasselas*.

In the Fourth Year, topics selected would include: the study of particular medieval plays; the poetry of Chaucer; the Theatre in Elizabethan times; sonnets of Sidney, and the allegory, of Spenser; the prose of such writers as Nashe, Deloney, North, Lyly, Elyot, and Ascham. Some plays of Shakespeare and some Jacobean drama would be read and discussed in detail.

From this it seems several light-years to the excellent teaching notes given to students in the same college:

Teaching English
(Secondary Modern and Junior Students)

(1) Where possible on Final Practice, try to use your English periods to compile a Class Magazine, or to prepare individual Anthologies, or recordings of visits, or local studies. (Try to get away from 'Composition'.)

(2) Where you can, link art and music, in any simple way, with poetry and literature.

(3) Try out dance drama, free drama, improvisation with movement, story themes; and try to explain to the children that this too is 'English'.

(4) Try always to encourage vivid and imaginative writing and speech, in verse and/or in prose, by reading good examples, in modern literature especially, and by fostering lively and 'real' discussion.

(5) When you read aloud, poetry or prose, from the Bible or from Shakespeare or from a history text-book, do your best to read in a clear voice which is sincerely expressive of the mood and the atmosphere of the words; forget to be self-conscious and be true to the author's intentions. Where you can, try to write something of your own to read aloud.

How can the aims behind both be reconciled?

There is no one solution in terms of one scheme: indeed, there might be a good deal to be said for doing what one hopes a head of department would do with his own teachers—that is for college of education lecturers to be told to take over a certain number of students, and to be given the injunction merely to teach them to read, to give them an adequate introduction to English literature, to enable them to articulate their experience of literature, and to practise creativity—each in his own way. No doubt this is how things work, in the best places—with discussion between tutors on the general aims and ground covered between them.

But such discussion needs to be based on a realistic sense of what does need doing. Here we need once more to summarise our positive aims. What is it a student teacher needs?

(1) An environment which is democratic in the sense that students are treated as responsible individuals, and on equal terms with the staff, as young developing adults, concerned with education in the sense of sharing the exploration of experience with them. Authoritarian set-ups, in which students are treated like boarding school pupils, harm personal development. The senior staff, while erecting no special aura round themselves, must at the same time be adult enough to take responsibility and to be independent adults. Thus they can help maintain a civilised environment in which individuals treat one another with respect.

(2) Experience of children should be pre-eminent, because it raises issues of personality development which tend to become forgotten or obscured in the young adult—buried, only to emerge again at awkward moments, such as when being 'tested' by a difficult child. At one college, teaching practice takes only one morning a week for 6 weeks in the first year; 3 whole weeks in the second year, and 6 weeks in the third year. This is simply not enough, surely?

Those students who get most out of their courses are noticeably those who have taught for a year or two before taking it. This seems ideal, and it is a pity this cannot become universal.

I hesitate to pronounce on actual times for teaching practice but I feel sure a whole term is valuable in the third year. Anything less does not give students enough initial introduction—the 6 weeks should come (perhaps as two periods of 3 weeks) in the first year, at the beginning. But short periods during term (such as 1 day a week) do give the students a chance to bring classroom experience freshly into Seminar discussion, for clarification. Other important experiences would seem to be:

(*a*) Visits to other kinds of educational establishments—E.S.N. schools, play-centres, child-therapy clinics, approved schools.

(*b*) The opportunity to observe other teachers teaching, of all kinds.

(*c*) The staff of training colleges should also as I have said teach in school for regular periods, in order to be able to report on their work, to continue to learn and develop, and to keep up with changes in the world of children (e.g. the vast new pressure from commercial entertainment).

(*d*) Special studies of children should be made: several colleges base their courses on child observation. Others have Study Practices—a student is introduced to two children, has to enter into a relationship with them, and to study them. He will take them out, teach them, explore their interests, their home background, their reading and writing, and everything he can discover about them.

(*e*) It may perhaps be added here that teachers in their probationary year need closer supervision, opportunities for seminar discussion of their work, and for refreshment in English. By this latter I mean courses in the reading and discussion of literature and of children's work.

(3) Teachers should have a sound experience of speech training. In my books on teaching English I have always given extensive details of exercises, games and projects for oral work: these must be practised by student teachers themselves. As the N.A.T.E. *Evidence* puts it:

Along with discussion and drama...speech education should play an important part...at the student's own level...The student must learn to read aloud, to tell stories, to discuss, to debate, to describe and to speak to a class effectively. In all these activities she should reach a high standard of fluent, well-mannered speech with good articulation. She must also learn the complementary art of sympathetic and intelligent listening...

It may be added that seminar work in itself, properly conducted, is one of the best trainings in the capacity to listen and to articulate clearly in speech.

(4) The English teacher should have a good deal of experience of creative work in the visual arts, in movement and drama, and in music, as well as in the writing of poetry and fiction—and discussion of creative work.

(5) There must be time to study children's writers (e.g. E. Nesbit, Lewis Carroll, Louisa Alcott, Mary Norton).

Having set out these needs, we may perhaps begin to see a different kind of approach to specifically 'English' studies emerging. We look first at young individuals of 19-20, in relationship to tutors of 30-60 on the one hand, and children of 5-15 on the other. We see the students' primary interest is in developing their own maturity, by creative work, by taking part in the lively creative, and sympathetic community of the college, and we see them beginning to explore the nature of children. In this especially (and more so if their practical experience of children comes earlier in their course) they begin to bring out their intuitive capacities to receive emanations from the children's own growing and in this to become flexible and to grow themselves.

Now we can see that we need in teacher training a process which reflects what I have insisted should be the dimension of English teaching for children: we begin not from a schematic diagram of 'Homer to T. S. Eliot', but from the young teacher's inward needs to explore his world, and to make sense of it—and find material from the body of literature which will foster this natural energy. When we investigate this, it will also reveal that there is not all that much distance between the 'inward child' of the student, and the child in the classroom, as we can see from student teachers' own poems.

I have already expressed my doubts over the time given to destructive modern writers. To endorse them even as mattering it seems to me is to begin a process which may make it harder to discover this natural self, and to put the study of literature in jeopardy, by establishing an atmosphere in which the expression of hate and fear, disguised as 'irony' or 'intelligence', will inhibit feeling and creativity. This is not to exclude works of maturity which contain destructiveness: indeed it is in them that the inner weak infantile self is discovered as the ultimately enduring reality—as in the 'endless humility' of *Four Quartets*, and in King Lear's

'I am a very foolish fond old man...'. What I do not mean is that one begins from callow expression and then 'brings the student on' (as one student recently put it) to 'Shelagh Delaney or Stan Barstow'. What we need to bring them on to should be sensitive and fine: to genuine creative minds capable of offering insight into weakness, and celebrating the human potentialities. If there is too little of this kind in modern writing, then we may leave it largely alone. Of course, privately students may read what they like: I am talking about what is endorsed by the syllabus.

But in any case what we *must* start from is the students' experience of a poem or a story: here there is no substitute if we are to do anything at all. Therefore I shall make my next point:

(6) The poems and stories selected for the initial seminars shall be of a very high quality as art, and those which, in the tutor's opinion, will raise the issues, terms, interests, attitudes which he wants to pursue.

And another:

(7) Some should be relevant to the interests of children: i.e. teachable in school. (Many training college courses include 'Eliot, Dylan Thomas, Hughes', but omit Edward Thomas and de la Mare—which is ludicrous and a failure to put students in touch with poets eminently relevant to the interests of children.)

The direction of our literary studies is governed now by themes, rather than 'periods' or 'forms'. And here, of course, the tutors need to discuss (*a*) how to step in with a formal lecture to 'tie things up' (and to demonstrate that they do have knowledge of their own which they can make available to the sudents), (*b*) how to make sure signposts are being set up, and the topography roughly surveyed, (*c*) how to make sure the exemplification of method is brought home, in terms of a self-conscious attention to method (what is it we have been doing?).

Now it remains only for me to sketch out some possible courses of direction an English approach of this kind might take, and then, simply, to list those works from which developments following themes, working in the way I have indicated, could draw material, over the three years. I ought also perhaps to add that I think 6 hours a week the minimum requirement for English (now including English 'method') for all students, and 1 whole day a week at least for 'advanced', 'main' and more for 'degree' students.

Since I have mentioned de la Mare, let me start with him. Here is a possible development for a first year.

(1) Discussion of some poems by Walter de la Mare: *Peacock Pie*, *The Charcoal Burner*, *Echo*, *Fear*, *Autumn* (two sessions).

(2) Melancholy and Nostalgia: Edward Thomas: *The New House*, *Melancholy* (one session).

(3) Nostalgia: D. H. Lawrence's *Piano*. Some of his poems about teaching (one session).

(4) *Sons and Lovers*: D. H. Lawrence (four sessions).

(5) Lecture: our attitudes to childhood.

(6) Talks by students in seminars:
>   on the *Nursery Rhymes* of the Opies;
>   on *Children's Lore*;
>   on children's writing and its themes.

(7) Attitudes to childhood in other times: *The Picture of Little T.C. in a Prospect of Flowers*, a poem by Andrew Marvell. Traherne, perhaps. If time, a poem by Henry Vaughan (one session). Perhaps a discussion of the childlike quality explored (and exploited) in hymns (Herbert to Mrs C. F. Alexander).

(8) *Oliver Twist* by Charles Dickens (three sessions).

(9) *Abraham and Isaac* from the Wakefield Mysteries (two sessions).

(10) *Portrait of the Artist as a Young Man* by James Joyce (four sessions).

Suggested Reading: The Country Child, Alison Uttley; *Father and Son*, Edmund Gosse; *A Memoir*, E.T.; *Huckleberry Finn*, Mark Twain; *The Mill on the Floss*, George Eliot; *The Secret Garden*, Mrs Frances Hodgson Burnett; *What Maisie Knew*, Henry James.

The tutor might add many possible fragments—Chaucer to his little son in his *Treatise on the Astrolabe*; Lewis Carroll; or he might turn to *The Winter's Tale* and its theme of fecundity and renewal. But what emerges is a *poetic* theme, and a sense of how many writers have (as we all do) used imagination to explore the nature of childhood and the relevance of some of its special qualities to adult perceptions. The next term one might begin again with de la Mare and take a quite different direction.

(1) de la Mare: discussion of a wider range of his verse in the light of F. R. Leavis's comments in *New Bearings in English Poetry*, about his attachment to magic. What is 'reality' and what kind of denials of reality do men make?

(2) Retrospection: Thomas Hardy on his loss: *Beeney Cliff*, *Boterel Castle*, *After a Journey*, *The Voice*.

(3) Stoicism in folksong: O *Waly Waly*, *The Seeds of Love*, *The Foggy Dew* (two sessions).

(4) The tragic view of life: *Odour of Chrysanthemums* (two sessions).

(5) The tragic view: *Heart of Darkness* by Joseph Conrad.

(6) The tragic view: *Nostromo* (ten sessions).

(7) The tragic view: *King Lear* and *Macbeth* (ten sessions).

Such exploration of poetic themes could be accompanied by material from music, opera, painting and ballet: there is surely no need to regard the area of 'Eng. Lit.' as so separate from the other arts? Indeed, the essence of such an approach should be that it should lead out, to considerations in which some students would inevitably become involved—questions of religion, philosophy and psychology, springing out of the theme. For instance, either of the above themes might suggest considerations of personal fulfilment and the significance of the personal life. This could well arouse considerations of what psychoanalysis, philosophy or ethics have contributed to our sense of potential fulfilment, and our sense of reality. A lecturer might well then take as a theme for his concluding discussion such paragraphs as these from Professor John Wisdom's *Philosophy and Psychoanalysis*. Professor Wisdom's themes are related to our concern to use the word as an instrument of insight, and thus of moral responsibility:*

Dr Waddington (in *Science and Ethics*) is anxious to increase our grasp of and emphasize the connection between ethics and psycho-analysis...The idea is still new though it has been put forward before to some extent, e.g. by H. V. Dicks when in *Clinical Studies in Psychopathology* (p. 109) he writes, 'Clearly the union of the opposites has been set as the highest goal of human achievement...a task to be fulfilled by the individual within himself—a process of psychological growth and unification—the resolution of conflict, to give it its modern name. The discovery of oneself, the finding of the centre from which we cannot err, of the "still, small voice", of the "Golden Flower"...of the thousand petalled lotus, etc., etc., by whatever name this precious self-realization and acceptance has been called—this is nothing less than the aim of psychotherapy, within the limits of the patients' powers...'

These remarks of H. V. Dick's have great importance for the question of moral education. From his own psychoanalysis Professor Wisdom discovered there was a distinction to be made between ethics (which is mental) and practical ethics (which is a matter of wholeness in living):

* See also E. Erikson, *Insight and Responsibility*.

By 'logical practice' I mean estimating the value of arguments, offering and accepting arguments and sifting our reactions to arguments. By 'ethical practice' I mean accepting and rejecting persons, acts and feelings, and the sifting of those acceptances and rejections. I mean the asking and answering of such questions as '*Is* Jack such a blackguard?' '*Ought* I to have done that?' 'Is it horrible of me to feel like I do when...?'

Creative English too can be a related process of self-discovery, of the discovery of personal potentialities and of those the world offers, and of 'ethical practice'.

Here I will end, having I hope done enough to suggest a whole different dimension of approach to the training of an English teacher, based on my own exploration of the nature of what goes on between teacher and child in the classroom—when it has bearing on the development of the 'thousand petalled lotus' and 'ethical practice', in the fullness of living.

19

ESSENTIAL RESOURCES

Let us assume that the college of education allows 4 hours a week for English for all students, and 6 hours a week for students taking English as a main subject, at advanced level, or for a B.Ed. It should be more: but the 'education' and 'psychology' interests are unlikely to be shifted. In good places 'education' and 'psychology' will become in part extensions of English: and some of the movement, mime and drama, speech, and other related work (such as folksong) will be done 'in' other subjects, or in collaboration with them. So let us consider 4 or 6 hours a week as a basic minimum for straightforward 'English' work.

This represents about 48 and 72 hours a term—say 50–75 hours. The second term should be given over more to teaching practice, and discussions of children and teaching: for our discussion of actual literature (including children's writing and students' own writing) we are left with 100 hours a year, or at most with 150 hours a year. This means 300 hours over the three years, or, say, 500 hours for 'advanced' students. This represents, roughly, the *seminar and lecture* time available, on actual works.

Now if, on the average, the seminar study of one poem takes half a seminar, a play about three seminars of 2 hours each, and a novel four seminars of 2 hours, we can see that our actual 'group possession' of works of literature may be seen as follows:

	hours
A. 90 poems	90
B. 12 plays	72
C. 24 novels	192
D. Lectures on critics, background, summaries, themes, perspectives	40
E. Children's writing, discussions of 60 pieces	60
F. Students taking seminars on their own work in English, say 15	30
G. Drama, mime, opera, free movement and dance	60
TOTAL	540

For E and F we can perhaps take some time from 'method' allocation under 'education', and G could be an 'extra' to the curriculum—an optional evening activity—though it should *not* be extracurricular. But certainly, there seems little time to waste on lectures on J. P. Donleavy or the latest 'brilliantly dirty' novelist. There is hardly time to put students in touch with the barest minimum of the more distinguished works of English literature.

Let us enumerate some, to fit that rough division, more or less, between 90 poems, 12 plays and 24 novels—and assume also that students will read in their own time. I emphasise that this is a list to *select* from. The works tutors select to pursue their 'themes' could be taken from the following. Column 1 is a list to select from for 'English for All' students; column 2 is for students taking advanced courses in English (the items from these two lists to be gone through in detailed discussion in seminars—really read and taught); column 3 is for students to read in their own time—starred titles being suggested for advanced students.

POETRY

1	2	3
Chaucer's *Prologue*; *The Wife of Bath's Prologue and Tale*	*The Nun's Priest's Tale*; *The Merchant's Tale*; *The Franklin's Tale*	*The Miller's Tale*; *Troylus and Criseyde*
		Piers Plowman (excerpts)
	Sir Gawaine and the Green Knight	Henryson: *The Testament of Creseid
Some early English lyrics—perhaps as songs set by modern composers (Britten, Mellers, Rubbra)	Poems by Dunbar, including *Lament for the Makers*	*The Two Married Women and the Widow (Dunbar)
Sir Thomas Wyatt: *The Appeal*, 'They flee from me...' etc.	More Wyatt (from the edition by Kenneth Muir)	
Songs by Dowland and Campion—with close discussion of the meaning and the relation of this to the music. (See Wilfrid Mellers, *Harmonious Meeting*)	Relate Dowland and Campion to folksong and Madrigals	Sir Walter Raleigh and Madrigals
	Spenser's *Prothalamion*	

Shakespeare's *Sonnets* (especially, 'When to the sessions...', 'That time of year...', 'They that have power to hurt...', 'When in the chronicle...', 'Let me not to the marriage...', 'Th'expense of spirit...', 'Poor soul the centre...'	Shakespeare's *Sonnets* complete. Songs from Shakespeare's plays. Songs from Ben Jonson	*Venus and Adonis*
John Donne: *The Ecstasy, The Apparition,* 'I wonder by my troth...'	Donne's *Songs and Sonets* complete	
Six poems by George Herbert including *Love, Virtue, The Pulley*	*The Temple* by George Herbert	*Metaphysical Poetry, Donne to Butler*★
Marvell's *Garden, Coy Mistress* and *Portrait of Little T.C. in a Prospect of Flowers*	Marvell's Poems, including the Horatian Ode, and *Dialogue Between the Resolved Soul and Created Pleasure*	
Traherne: a few poems. (See comment by Marion Milner, *On not Being Able to Paint*)		
	King's *Exequy*	
Henry Vaughan: *The Retreat, Quickness*	Some Vaughan and Crashaw, including Crashaw's *Hymn to Saint Theresa*	
Milton: *Lycidas*		★*Paradise Lost*
	Bunyan's Hymns	
Border Ballads: *Sir Patrick Spens, Clerk Saunders, Barbara Allen, The Lyke-Wake Dirge*	(Link the Border Ballads with folksong)	Ballads and folksong: a special study

Pope: *Moral Essays on The Use of Riches*   *The Rape of the Lock*

Johnson: *London* and *The Vanity of Human Wishes*

William Collins: *Ode to Evening*

Christopher Smart: *Song to David, To His Cat Jeffery*

Crabbe: *The Parish Register* I and III, *The Lover's Journey*   Crabbe: *Peter Grimes, The Frank Courtship, Procrastination*

Blake: *The Sunflower, The Poison Tree, The Tiger,* and other *Songs of Experience*   Blake: *Songs of Innocence and Experience,* and other poems

Burns: a few poems

Wordsworth: *The Lucy Poems, Upon Westminster Bridge, Surpris'd by Joy, The Solitary Reaper, Margaret*   *Michael, Lines on the Simplon Pass, The Leech Gatherer*   ★*The Prelude, Intimations of Immortality, Tintern Abbey*

Coleridge: *The Ancient Mariner, Frost at Midnight*   *Kubla Khan*

Shelley: *Ode to the West Wind*

Clare: *I Am, The Pettichap's Nest.* (A selection from Clare as in *Iron, Honey, Gold* perhaps)   Clare: a full study (20–30 poems)

Keats: *Ode to Autumn, Ode to a Nightingale, Ode on a Grecian Urn, Ode to Melancholy*   *The Eve of St Agnes, The Revised Hyperion, La Belle Dame Sans Merci, On First Looking into Chapman's Homer*

Browning: *Meeting at Night, The Bishop Orders His Tomb, In a Gondola*

	Tennyson: *Summer Night*, some of *In Memoriam*	
Emily Brontë: *Deep in the earth, Often rebuked*	Emily Brontë: *Collected Poems*	
	Christina Rossetti: a selection, as in *Iron, Honey, Gold*. Emily Dickinson: some poems from *Selected Poems*	
Hardy: *The Going, The Voice, After a Journey*	*Veteris Vestigia Flammae*, and other poems	
Hopkins: *The Windhover, The Leaden Echo, That Nature is a Heraclitean Fire, Pied Beauty*	Hopkins: a full study	
Yeats: *The Magi, Sailing to Byzantium, A Prayer for my Daughter, Among School Children, Byzantium*	Yeats: *The Falling of the Leaves, The Sorrow of Love, The Song of Wandering Aengus, He gives his beloved, The folly of being comforted, Adam's Curse, The fascination of what's difficult, No second Troy, Two Songs of a fool, The Second Coming, A Dialogue of Self and Soul*	Yeats's *Collected Poems* (see also under drama)
Ezra Pound: A selection (as in *Iron, Honey, Gold*)	*Hugh Selwyn Mauberley* and poems from the *Lustra* Section of *Collected Poems*	
T. S. Eliot: *The Waste Land, Preludes, Ash Wednesday, Journey of the Magi, Marina Animula*	*Four Quartets, Gerontion*	

D. H. Lawrence: *Love on the Farm, The Colliers' Wife, The Best of School, End of Another Home Holiday, Baby Running Barefoot, Baby Asleep After Pain, Last Lesson of the Afternoon, School on the Outskirts, A Snowy Day in School, Reminder, Last Words to Miriam, Release, Kisses in the Rain, Snapdragon, Man and Bat, Snake, Mountain Lion, Humming Bird, Bavarian Gentians, Peach, Fish*

Look! We Have Come Through!

*The Complete Poems

Poets of the 1914–18 war: Isaac Rosenberg: *Dead Man's Dump, Break of Day in the Trenches, Returning we hear the larks, Daughters of War*

Rosenberg: *Collected Poems*

Wilfrid Owen: *Miners, Anthem for Doomed Youth*

Wilfrid Owen: *Collected Poems*

Edgell Rickword: *The Soldier Addresses his Body, Winter Warfare, Trench Poets*

Some poems by:
Robert Graves,
Herbert Read,
Siegfried Sassoon

Edward Thomas: *The Owl, Thaw, Melancholy, Cock Crow, Old Man, The New House, Digging*

Edward Thomas: *Collected Poems*

Some Robert Frost (*Wall, Storm, Fear*)

Robert Frost: *Collected Poems*

A selection of modern poets: William Empson, W. H. Auden, Cecil Day Lewis, Andrew Young, Louis Mac-Neice, the half dozen or so 'true' poems of Dylan Thomas, Edwin Muir, Kathleen Raine, Henry Reed, Guy Butler (South African poet), Ted Hughes, Bernard Stevens, Theodore Roethke

Walter de la Mare: *Tom's Angel, Solitude, Futility, Echo, The Argument, The House, An Epitaph, Alone, Peeping Tom*

Walter de la Mare: *Peacock Pie*

Translations from the Chinese by Arthur Waley: Anon; *The Orphan, The Sick Wife.* (And other Chinese poems as selected in *Iron, Honey, Gold*)†

170 Chinese Poems, translated by Waley†

Hymns: *The Cambridge Hymnal*

Penguin Modern Poets as published*
Oxford Nursery Rhyme Book. The Lore and Language of School-children (Iona and Peter Opie)

The first column represents what seems to me the minimum 'possession' of English poetry necessary for an English teacher: yet for

* Colleges should join the Poetry Book Society to receive modern verse regularly.
† An edition of Chinese Poetry in translation for school use is being prepared by the present writer.

genuine reading and discussion the number of hours' seminars required I reckon to be about 150 hours, and for the advanced students (column 1 plus column 2) 400 hours altogether, to cover the complete list in discussion, plus 200 hours of private reading. This work must be done before spending effort on ephemeral modern verse, or Sunday-newspaper fashionable 'masterpieces'. Students should also make a full study of the available school anthologies.

Much the same applies to drama: no doubt in their own time students can find excitement in 'Fry, Ionesco, Bolt, Osborne, Wesker'. But there is a profound need for them to acquaint themselves first with the highlights of English drama, and this takes time, if it is to be done properly:

DRAMA

1	2	3
The Towneley Second Shepherd's Play of the Nativity	The Towneley Abraham and Isaac	The Wakefield Mystery Plays as a whole
		David Linsay: Ane Satire of the Three Estatis
		Sophocles's Antigone★
	Milton: Comus	
Marlowe: Dr Faustus	Marlowe: Edward II	
Shakespeare: Richard II, Twelfth Night, Henry IV, Macbeth, The Tempest, Julius Caesar	Shakespeare: King Lear, Coriolanus, The Winter's Tale	Shakespeare★: Troilus and Cressida, Richard III, Romeo and Juliet, Anthony and Cleopatra, Henry IV Part 2, Henry V, Hamlet, Othello
Dekker: Shoemaker's Holiday		
Tourneur: The Revenger's Tragedy	Webster: The Duchess of Malfi	
Jonson: The Alchemist	Jonson: Volpone	

Ibsen: *Peer Gynt, Ghosts, An Enemy of the People*

Synge: *The Playboy of the Western World*

Yeats: *Calvary, Resurrection*

Yeats: *On Baile's Strand*

T. S. Eliot: *Murder in the Cathedral*

T. S. Eliot: *Sweeney Agonistes, The Family Reunion*

Ezra Pound: *The Women of Trachis* (from Sophocles)

John Arden: *Sergeant Musgrave's Dance*

Japanese Noh Plays: *The Damask Drum*, etc.

Some modern drama: see the series edited by Michael Marland for Blackie

Here again the *merest* acquaintance with good drama suggests something like 80 hours for the basic English course, and another 100 hours for 'advanced' students. Besides the study of these texts, students should be acquainted with some of the experiments in modern drama-with-music such as *Let's Make an Opera, Noye's Fludde* and *Peter Grimes* (Britten); *Porgy and Bess* (Gershwin); *L'Enfant et les Sortilèges* (Ravel); *Der Mond* (Orff); *The Tender Land* and *The Second Hurricane* (Aaron Copland). Several of these are 'school' operas.

Modern drama can be left mostly to the students' own leisure pursuits (and to enthusiastic drama tutors!).

With prose we shall have to be even more stringent. Starred titles here represent works probably only possible for four-year students.

PROSE

1	2	3
Passages from Ecclesiastes, some psalms	The Book of Job, Revelation (excerpts), Judith (Apocrypha)	The Authorised Version of the Bible (see choice of passages in *English for Maturity*)

Sir Walter Raleigh: *Letters*		
	Donne's Sermons (selected)	
Nashe, from *The Unfortunate Traveller* (extracts)		
	Bunyan: *Pilgrim's Progress*. Swift: *A Voyage to Lilliput*	Defoe: *Robinson Crusoe*. Fielding: *Tom Jones*
	Johnson: *Life of Cowley*. Halifax: *The Character of a Trimmer* (extracts). Cobbett: *Rural Rides* (extracts)	
Jane Austen: *Sense and Sensibility*	*Emma, Persuasion*	*Pride and Prejudice*
Emily Brontë: *Wuthering Heights*		
Dickens: *Oliver Twist, Bleak House*	*Little Dorrit, Hard Times, Dombey and Son*★	*Martin Chuzzlewit, David Copperfield, Great Expectations*
George Eliot: *The Mill on the Floss, Felix Holt* or *Middlemarch*	*Daniel Deronda*★ (or *Gwendolen Harleth*— i.e. the shortened version recommended by F. R. Leavis)	*Silas Marner, Adam Bede*
Keats' *letters*		
Thomas Hardy: *The Woodlanders*	*The Trumpet Major, Jude the Obscure*★	*Tess of the D'Urbervilles*
Henry James: *What Maisie Knew*	*The Bostonians, Portrait of a Lady*★	*The Awkward Age*
Joseph Conrad: *Heart of Darkness, Typhoon, The Secret Agent, The shadow-line*	*Nostromo*★, *Youth, The End of the Tether*	*Under Western Eyes, Chance*
Mark Twain: *Huckleberry Finn*	*Puddn'head Wilson*	Mark Twain: complete works

D. H. Lawrence: *Sons and Lovers, Odour of Chrysanthemums, The Prussian Officer, The White Stockings, The Captain's Doll* — *Women in Love, The Rainbow★, Collected Short Stories★* — *Kangaroo, The Lost Girl, Aaron's Rod*

E. M. Forster: *Where Angels Fear to Tread* — *A Passage to India* — *Howard's End★*

James Joyce: *Dubliners* (e.g. *Grace, The Dead*) — *Ulysses★*

T. F. Powys: *God's Eyes A-Twinkle* — *Mr Weston's Good Wine* — *Mr Tasker's Gods*

George Douglas Brown: *The House with the Green Shutters*

Robert Graves: *Goodbye to All That*

F. Sturgis: *Belchamber★*

Virginia Woolf: *To the Lighthouse*

Edith Wharton: *The House of Mirth*

Arthur Koestler: *Darkness at Noon*

Maxim Gorki: *Childhood* — Tolstoy: *War and Peace★*. Dostoevsky: *Crime and Punishment* — Tolstoy: *Anna Karenina★*

L. H. Myers: *The Root and the Flower*

Ernest Hemingway: *'Nick' Stories, A Farewell to Arms* — Scott Fitzgerald: *The Great Gatsby* — George Orwell: *Animal Farm, 1984*

Here again, there is hardly time to find one's way about the *essential* works of distinction and read a selection from them. The above lists would require, for seminar treatment, 140 hours from the 'basic' students, and 260 hours for 'advanced' students and another 80 for B.Ed. or fourth-year students. How can there be time for 'Angus Wilson, Iris Murdoch, Kingsley Amis, C. P. Snow'? Of course, students can be given lists of the more interesting modern novels (Doris Lessings' early books, Raymond Williams' *Border Country*) and poets (Ted

Hughes, Roethke). And then some useful books for children such as Emma Smith's *Maiden's Trip*; Richard Church's *The Cave*; Edmund Gosse's *Father and Son*; Richard Hughes's *High Wind in Jamaica*; Thomas Hardy's *Our Exploits at West Poley*. But I have only given a sketch of barest minimum requirements, only too aware that I have left much out and these total as follows:

	hours		hours
Basic poetry	150	Advanced poetry	400
Basic drama	80	Advanced drama	180
Basic prose	140	Advanced prose	260–340
TOTAL:	370	TOTAL:	840–920

But, in fact, we probably only have 300 hours with 'basic' students and 500 with 'advanced' students! So, the tutor has to select drastically from such a list of 'best works'. To this must be added the whole problem of putting the student English teacher in touch with children's literature, with popular culture, folksong, and other related topics.

Selecting even from some such selections the English tutor must seek to put future English teachers in touch with creative language used at its finest: only thus can they take the 'feel' of touchstone by which to judge the creative use of language by their pupils.

This purpose is not served by substituting the history of Eng. Lit. from the outside, as happens for exam purposes. It is not served by substituting linguistics for response to the meaning of words in the whole experience of word-art. It is not served by giving the course over to the values of Sunday-newspaper journalism, and its bogus intellectualism. Nowadays there is almost a cliché substitute for proper literature studies in the college of education consisting of a course centred on such books as J. D. Salinger's *The Catcher in the Rye* and Golding's *Lord of the Flies* with little sense of where to go after, nor of how limited these books are.

This is not the place to make a critical analysis of these. Anyone who believes them to be a realistic portrait of childhood must have forgotten his own. Salinger seems to me to be imprisoned in his own idiom, with the consequent effect of implying that a sensibility can be limited to the 'type' self of adolescence more than a sensibility ever is, by adolescence or cultural predicament: for his own purposes he limits his boy hero's sensitivity. The idiom and the protagonist created in it are tools for the attempt to convey a depressively sentimental, negative and stand-aside attitude to life. This book is itself a symptom of the 'taboo on tenderness', because it implies that the youth's inner self can summon neither strength, resources, creativity nor beauty—and cannot break out of its environment any more than the writer can break out of his defensive throwaway idiom: one can only be an 'outsider'. Golding's book implies a denial of the reparative impulse and so of human reality: it is really no more than a thriller, disguised by portentous gestures. The prose in which it is written is essentially magazine prose whose gestures disguise the failure to discover the truth that 'civilisation is created anew in each child'—in a positive sense.

Such issues as these books raise, in a blunt-fingered way, may be raised in a much finer way by a story or novel by Lawrence or Mark Twain, whose *Huckleberry Finn* is, as Leavis says, an exploration of the whole nature of civilisation. Indeed we might endorse the use of Salinger if he were used in comparison with Mark Twain, to show how Twain's idiom is a recreation of the questioning vitality of childhood, while Salinger's is the confining of it within the limitations of a modish adult cliché, and the attitudes of sentimental hopelessness that go with it. The way teacher training, at its more 'progressive', tends to stick at such works is a mark of some Philistinism at the roots of our cultural life.

In order to thrash out such problems perhaps what is needed, still further in the background, is a college of education at which the staffs of colleges of education themselves can be trained and refreshed, in which such issues could be fought out, as they need to be fought out, at adult level, and fresh creative standards developed. The patchwork nature of college of education courses, in so many places apparently missing the real work altogether, suggests that the trouble is ignorance, and that too little relevant thinking and planning has been done. This book is a first gesture towards thoroughly needed reform.

APPENDIX A

THE LITERATURE OF CHILDHOOD

Such a study needs perhaps to begin with a look at those books first published in the eighteen-seventies and eighteen-eighties which are still so successful—Ralph Caldecott's books (1870), Walter Crane's *The Baby's Opera* (1876), Kate Greenaway's *A—Apple-pie* (1886) and *Marigold Garden* (1885), and especially the *Little Black Sambo* and *Mingo* books of Helen Bannerman, with their terrifying, but to a child satisfying, stories of annihilation, and triumph over disintegration. (They were first published in 1899 and are still selling 25,000 a year each.) Then to books for four-year-olds by Beatrix Potter, with her world of animal badness and goodness, threat and comfort, with many excitingly out-of-the-way words (like 'soporific') thrown in. Students must be prepared to study the early world of childhood, for to understand the phantasy of older children we must study their earlier imaginative needs.

Here students would need to study how language is full of magic to a child from the beginning and how children may possess a word like 'cautiously' or 'soporific' and perform all kinds of inward magic with it, for a long time, before they have any idea what it means. The study of children's books therefore helps lead into the study of attitudes to children, and to theories of child development.

Students should study processes of identification with animals in phantasies, in which children 'become' the protagonist and his own struggle between good and bad. From Babar and Beatrix Potter students can graduate to *Little Tim* and *Paul the Hero of the Fire*: they should also be encouraged to ask what makes such books of more value than Enid Blyton. To such studies we may add the *Oxford Book of Poetry for Juniors*, the *Puffin Nursery Rhyme* book, the Opies' *Oxford Book of Nursery Rhymes*, and the whole range of child lore. This connects with the study of folksong.

Students could begin with the *Orlando* books—and debate the contrast in them between the rather chi-chi writing and the much more interesting pictures of the cat's predicaments. Even though students are going to work with older children, they should study the literature of early childhood to try to discover why children are so fascinated by the discomforts of Hans Andersen's or Grimms' Fairy Tales. What is good about *Winnie the Pooh*? Where does Kipling come in? What is the value of *English Fables* retold by James Reeves, Alison Uttley's *Tales of Sam Pig*, or *Brock the Badger*? After this stage more adult standards of judgement are relevant, and students should certainly discuss the great classics of children's reading: E. Nesbit is, for instance, a delightful writer. *Alice in Wonderland* and *Huckleberry Finn* are essentials for discussion. (There is a valuable, if not critically consistent, series of monographs on children's writers published by The Bodley Head.)

The following is a list of the books a fairly bright child should have read by the time he or she is fourteen: the student teacher should surely have read them all and should make a study of at least one author among them.

Louisa Alcott. *Little Women*, 1868; *Good Wives*, 1871; *Little Men*, 1871; *Jo's Boys*, 1886.

Hans Andersen. *Forty-two Stories* (Faber); *Hans Andersen's Fairy Tales: A Selection* (Oxford).

Edward Ardizzone. *Little Tim and the Brave Sea Captain*, *Lucy Brown and Mr Grimes*; *Tim and Lucy go to Sea*; *Tim and Charlotte*; and so on (Oxford).

Helen Bannerman. *Little Black Sambo*; *Little Black Mingo*; *Little Black Quasha*; *Little Black Quibba*; *Little Black Bobtail* (Chatto and Windus).

Paul Berna. *A Hundred Million Francs*; *The Street Musician* (Bodley Head).

Frances Browne. *Grannie's Wonderful Chair*, 1857 (Blackie).

Frances Hodgson Burnett. *The Secret Garden*, 1911 (Heinemann); Compare *Little Lord Fauntleroy*.

Lewis Carroll. *Alice in Wonderland*, 1865; *Alice Through the Looking Glass*, 1872.

Erskine Childers. *The Riddle of the Sands*.

Richard Church. *The Cave*, 1963 (Dent).

Susan Coolidge. *What Katy Did*, 1872; *What Katy Did at School*, 1873; *What Katy Did Next*, 1886 (Blackie).

Jean de Brunhoff. *The Story of Babar the Little Elephant*, and many others.

Meindert de Jong. *The Wheel on the School*, 1954 (Lutterworth).

Walter de la Mare. *Collected Stories for Children*, 1947 (Faber).

Elizabeth Enright. *Thimble Summer*; *Then there were Five*, etc. (Heinemann).

David Garnett. *Pochahontas*, 1933 (Chatto and Windus).

Eve Garnett. *The Family from One-End Street*; *Further Adventures of the Family from One End Street* (Heinemann).

Kenneth Grahame. *The Wind in the Willows*, 1905 (Macmillan); also *Toad of Toad Hall*, the same book made into a play, by A. A. Milne.

Kate Greenaway. *Under the Window*, 1879; *Marigold Garden*, 1885 (Warne, reissued).

Frederick Grice. *The Bonny Pit Laddie*, 1960 (Oxford).

Grimm, Jacob and W. K. *Grimms' Fairy Tales* (Oxford).

Kathleen Hale. *Orlando the Marmalade Cat*; and many others (Murray).

Erich Kästner. *Emil and the Detectives*, 1931; *The Flying Classroom*, 1933; *Emil and the Three Twins*, 1949; *The Thirty fifth of May*, 1949; *Lottie and Lisa*, 1949 (Cape).

Rudyard Kipling. *The Just So Stories*; *The Jungle Book*.

Phyllis Krasilowoky. *The Cow Who Fell into the Canal*, 1958.

Barbara Leigh. *Carbonel*, 1955.

C. Day Lewis. *The Otterbury Incident*, 1958 (Putnam).

Kathleen Lines. *The Ten Minute Story Book* (Oxford).

Marjorie Lloyd. *Fell Farm Holiday*; and other Fell Farm stories.

George Macdonald. *At the Back of the North Wind*, 1871; *The Princess and the Goblin*, 1872; *The Princess and Curdie*, 1883 (Dent).

A. A. Milne. *Winnie the Pooh*, 1926; *The House at Pooh Corner*, 1928 (Methuen).

E. Nesbit. *The Railway Children*.

Mary Norton. *The Borrowers*, 1952; *The Borrowers Afield*, 1955; *The Borrowers Afloat*, 1959; etc. (Dent).

Iona and Peter Opie. *The Oxford Book of Nursery Rhymes*; *The Language and Lore of Schoolchildren* (Oxford).

Philippa Pearce. *Tom's Midnight Garden*, 1958 (Oxford).

Gene Stratton Porter. *A Girl of the Limberlost*.

Beatrix Potter. *The Tale of Peter Rabbit*; *The Tale of Jemima Puddle-Duck*; *The Tale of the Flopsy Bunnies*; and many others, including *Jérémie Pêche-à-la-Ligne*; *Nanes Berda Bynni*; and *Die Geschichte Der Nasenfamilie Plumps* (Warne).

Arthur Ransome. *Swallows and Amazons*, 1930, etc. (Cape).

James Reeves. *English Fables and Fairy Stories* retold by James Reeves (Oxford); *Pigeons and Princesses* (Heinemann); *A Golden Land*; *Stories, Poems, Songs Old and New* (Constable).

H. A. Rey. *See the Circus*; *Where's My Baby?* (Flap Books, Chatto and Windus).

Anna Sewell. *Black Beauty*, 1877 (Dent).

R. L. Stevenson. *Treasure Island*.

Mark Twain. *Huckleberry Finn*; *The Adventures of Tom Sawyer*; *Pudd'nhead Wilson*; *Roughing It*; *A Yankee at the Court of King Arthur*; *The Prince and the Pauper*.

Alison Uttley. *Adventures of Sam Pig*, etc. (Faber).
Rutger van der Loeff. *Avalanche*, 1951 (University of London).
Henry Williamson. *Tarka the Otter*; *Salar the Salmon*.

POETRY EDITIONS FOR CHILDREN

Lavender's Blue, ed. Kathleen Lines (Oxford University Press); *Stars and Primroses*, ed. M. C. Green (Bodley Head); *Magic Lanterns*, ed. M. C. Green (Bodley Head); *Come Hither*, ed. Walter de la Mare (Faber and Faber); *The Puffin Book of Verse*, ed. Eleanore Graham; *The Merry Go Round*, ed. James Reeves (Heinemann).

BOOKS ABOUT CHILDREN'S READING

Four to Fourteen, Kathleen Lines (Cambridge, for the National Book League, 1956); *Tales out of School*, Geoffrey Trease (Heinemann, 1949); *About Books for Children*, Dorothy Neal White (Oxford, 1949); *Books Before Five*, Dorothy Neal White (Oxford); *Intent Upon Reading*, Marjorie Fisher (Lutterworth, 1954); *Books at Bedtime*, by David Holbrook (A Newman Neame Take Home Book, 1962).

To these lists we must add a list of books from which the tutor or student teacher can select examples of children's writing for seminar discussion, such as:

Young Writers, Young Readers, ed. Boris Ford; *Let the Children Write*, Margaret Langdon; *Life Through Young Eyes*, Dolphin; *An Experiment in Education*, Sybil Marshall; *The Excitement of Writing*, A. B. Clegg; *The Keen Edge*, Jack Beckett; and *Coming into Their Own*, Marjorie Hourd.

APPENDIX B

A NOTE ON 'WHOLENESS'

I often use the phrases 'whole persons' or 'experience in whole terms'. I am adding this note, to seek to give substance to such expressions, should the phrases be challenged.

A poet or novelist tries to deal with the whole of experience. When we write books of theory by contrast we make abstractions from experience. These abstractions may refer to the objective world (as in a book on biology) or to the subjective world (as in a book on the structure of personality or—say—problems of conscience). The problem in education seems to me to keep abstract discussion to heel, by continual reference to whole experience and what goes on in whole persons. Many troubles arise from the way in which educational training and organisation tend to be governed by abstract principles derived from the measurement of partial functions—whose character and existence in any case depends upon subjective factors which are not measurable, and are not taken account of by mensuration.

'Whole' experience, then, includes both subjective and objective reality. Here the first problem is to secure the acceptance of the existence of inner reality. I find students have some difficulty in recognising this world of the 'unknown self' or 'inner being'. Here a psychological statement of the nature of 'inner reality' may help:

The mind or psyche has a reality of its own, separate and distinct from the reality of the outer material world. It has its own enduring and not easily alterable organization. The psyche has, one might almost say, a kind of solid substantiality of its own which we cannot alter at will, and which we have to begin by accepting and respecting. Thus, we cannot ourselves *feel* differently from the ways in which we discover that we do feel. We do not choose what we shall feel, we simply discover that we are feeling that way, even if we have some choice in what we do about its expression. Our feelings are instantaneous, spontaneous, and at first unconscious reactions which reveal the psychic reality of our make-up. At any given moment we are what we are, and we can become different only by slow processes of growth. All this is equally true of other people who cannot, just because we wish it, suddenly become different from what they are. Psychic reality, the inner constitution and organisation of each individual mind, is highly resistant to change, and goes its own way much less influenced by the outer world than we like to think.

Our conscious mental operations do not convey the full force of this stubborn durability of psychic reality, since it is relatively easy to change our ideas, to alter our decisions, to vary our pursuits and interests, and so on; but we can do all that without becoming very different basically as persons. Our mental life appears to be a freely adaptable instrument of our practical purposes in the outer world, as no doubt it should be. The closer, however, we get to matters involving the hidden pressures of emotions, the more do we recognise the apparent intractability of psychic reality. The infatuated man cannot subdue his infatuation, the person who worries cannot stop worrying, the hyperconscientious person who works to death cannot relax, the man with an irrational hate cannot conquer his dislike, the sufferer from bad dreams cannot decide not to have them. This is conspicuously the case with neurotic persons, who manifest a marked helplessness towards their own psychic reality and emotional life...*

Now one of the ways in which we can seem to triumph over this intractable inner world, is by exercising the abstracting intellect over it. We find it painful and complex to take in the whole of experience, so we ratiocinate and turn everything we can into an abstraction by ideation. These mind-constructions can then *seem* to be the world: what we fail to see is what our system has left out. By devising logical and rational schemes we can seem to tie up our experience in manageable ways: we can seem to be able to subject the disturbing and intractable subjective to the objective.

In fact, all we have really done is to ignore whole areas of our inner world, or our whole experience: we have pretended we can exist and manage in a split form. In this is the source of what T. S. Eliot called 'the dissociation of sensibility': in our era, since the growth of science and technology from the seventeenth century, men have increasingly divided themselves between the world of feeling and the world of action and outward effectiveness. And they have sought to solve their problems of 'inner goodness' or psychic integration, less and less by creativity, and more and more by 'outer goodness'—by material developments and by seeking to gain power over the external world.

This is not the place to pursue this theme. But perhaps I may refer the reader to the very close and fascinating exploration of the problems of subjective and objective knowledge, from the point of view of a biologist and educationist, by Marion Milner, in her book *On Not Being*

* H. Guntrip, *Personality Structure and Human Interaction*, (Hogarth Press, 1961), pp. 218-19.

Able to Paint. Her book is a record of the search of a biologist and educational research worker for her own creativity, while undergoing psychoanalysis. Gradually, she became aware of the nature of *being* in a whole sense, of the capacity to embrace both objective and subjective reality, and to discover the self as a whole being in relation to these worlds as a whole. As an educationist she concluded:

Having made such a discovery about a different way of being, the question arose, could not educational theory and practice somehow find out more about it and come to make a more deliberate allowance for it; rather than concentrating so much on the way that stands apart and only tries to give an objective detached account of what it sees? In fact, I suspected that if education did manage to do this, one of the results would be not a lessening of the objective powers but a strengthening...

The discovery and explanation of the self and the subjective world, she argues, would lead children to find touch with others, and so to greater social capacities. And the effect would be to lessen the amount of hate in education—that is, the desire of the primitive infant in us to subdue the subjective world by sheer force of will and the urge to incorporate what it could not allow to exist:

...was it not possible that orthodox educational methods increased the hate resulting from the primary disillusion [i.e. when one was first obliged to recognise that one could not make and command the world by phantasy, but was a real weak helpless creature in a real world] not only by not giving enough scope for the aesthetic way of transcending that hate, but also by not giving enough scope for the social way?

In fact it has been increasingly demonstrated to what extent creative opportunities in school help children to overcome 'primitive hate' in themselves, and to find order, beauty and self-possession. (Significantly, Mrs Milner has a footnote at this point on Marjorie Hourd.) 'Psychic creativity' in school can be the basis of personality development, and of greater sociability. Here in the present work I discuss once more in a different context some of the work done by children whose histories are given in *English for the Rejected* and *The Secret Places*, where I followed children in their employment of creative disciplines to come to terms with themselves and to assert themselves, towards beauty and aspiration. They came to self-realisation through exploring their 'whole' experience, and also to more effective 'practical' (or objective) capacities.

BIBLIOGRAPHY

EDUCATION FOR TEACHING

Education and the Training of Teachers: A Plea for the Education of Teachers as Persons, P. Gurrey (Longmans).

THE DEVELOPMENT OF PERSONALITY AND THE
NATURE OF EDUCATION

Feeling and Perception in Young Children, Len Chaloner (Tavistock).
The Origins of Love and Hate, Ian D. Suttie (Penguin Books).
On Not Being Able to Paint, Marion Milner (Heinemann).
Intellectual Growth in Young Children, Susan Isaacs (Routledge).
Introduction to the Work of Melanie Klein, H. Segal (Heinemann).
Envy and Gratitude, Melanie Klein (Tavistock).
Love, Hate and Reparation, Melanie Klein and Joan Riviera (Tavistock).
Our Adult Society and its Roots in Infancy, Melanie Klein (Tavistock).
The Child and the Outside World, D. W. Winnicott (Tavistock).
The Child and the Family, D. W. Winnicott (Tavistock).
The Family and Individual Development, D. W. Winnicott (Tavistock).
Collected Papers, D. W. Winnicott (Tavistock).
The Maturational Processes and the Facilitating Environment, D. W. Winnicott (Hogarth Press).
Excerpts from Winnicott in *The Child, the Family and the Outside World* (Penguin Books).
Personality Structure and Human Interaction, H. Guntrip (Hogarth Press).
Psycho-analytical Studies of the Personality, W. R. D. Fairbairn (Tavistock).

TEACHING ENGLISH

English for Maturity, David Holbrook (Cambridge).
English for the Rejected, David Holbrook (Cambridge).
The Secret Places, David Holbrook (Methuen).
Teaching English, J. H. Walsh (Heinemann).
English in Education, Denys Thompson and Brian Jackson (Chatto and Windus).
The Excitement of Writing, A. B. Clegg (Chatto and Windus).
Let the Children Write, Margaret Langdon (Longmans).
An Experiment in Education, Sybil Marshall (Cambridge).
Free Writing, Dora Pym (London University Press).
The Keen Edge, Jack Beckett (Blackie).
Topics in English, Geoffrey Summerfield (Batsford).

Coming into Their Own, M. L. Hourd and G. E. Cooper (Heinemann).
The Education of the Poetic Spirit, M. L. Hourd (Heinemann).
Some Emotional Aspects of Learning, M. L. Hourd (Heinemann).
Freedom and Authority in Education, G. H. Bantock (Faber).
Education in an Industrial Society, G. H. Bantock (Faber).
Education and Values, G. H. Bantock (Faber).
A Question of Living, R. F. Mackenzie (Collins).
English in the C.S.E., A Working Party of Teachers (Cambridge).
The Disappearing Dais, Frank Whitehead (Chatto and Windus).

For books on subdivisions of English Teaching (poetry, folksong, reading, language work, drama), see the bibliography of *English for Maturity*.

ASPECTS OF POPULAR CULTURE

GENERAL

Discrimination and Popular Culture, ed. Denys Thompson (Penguin Books).
The Uses of Literacy, Richard Hoggart (Chatto and Windus).
The Popular Arts, Stuart Hall and Paddy Whanel (Hutchinson).
The Affluent Society, J. K. Galbraith (Penguin Books).
Communications, Raymond Williams (Penguin Books).
Understanding the Mass Media, Nicholas Tucker (Cambridge).
Popular Culture and Personal Responsibility: A Study Outline (N.U.T.).

ADVERTISING

The Voice of Civilisation, Denys Thompson (Muller).
Your Money's Worth, Elizabeth Gundrey (Penguin Books).
The Economics of Advertising, F. P. Bishop (Robert Hale).
Advertising and Economic Theory, E. A. Lever (Robert Hale).
The Ethics of Advertising, E. A. Lever (Robert Hale).
Advertising in Modern Life, J. Gloag (Heinemann).
Truth in Advertising, W. Weir.
The Affluent Sheep, Robert Millar (Longmans).
Advertising: A New Approach, Walter Taplin (Hutchinson).
Advertising, Alan Hancock (Longmans).
The Shocking History of Advertising, E. S. Turner (Michael Joseph).
Advertising: A General Introduction, Robert S. Caplin, F.I.P.A. (Business Publications Ltd.).
The Hidden Persuaders, Vance Packard (Penguin Books).
The Image, Daniel Boorstin (Penguin Books).
The Mechanical Bride, Marshall McLuhan (Vanguard).

BIBLIOGRAPHY

COMICS AND TELEVISION

Seduction of the Innocent, Frederick Wertham (Museum Press).

Television and the Child, Dr H. Himmelweit, A. Oppenheim and P. Vince (Oxford).

Television in the Lives of our Children, Schramm, Lyle and Parker (Oxford).

LINGUISTICS

Problems and Principles in Language Study, David Aberchombie (Longmans, 1956).

Speech and the Tongues of Men, J. R. Firth (Oxford, repr. 1966).

The English Language, An Introduction, W. Nelson Francis (Norton, U.S.A., 1966).

Linguistics and English Grammar, Henry A. Gleason (Holt, Rinehart, Winston, U.S.A., 1965).

The Use of English, Randolph Quirk, (Longmans, 1962).

Language, Edward Sapir (American paperback).

Modern English Structure, Barbara M. H. Strang (Edward Arnold, 1962).

Transformational Grammar and the Teacher of English, Owen Thomas (Holt, Rinehart, Winston, U.S.A., 1965).

INDEX

The lists of authors and works in chapter 19 and Appendix A are not included in this index.

PB⊘-9041